*Laser Welding*

# Laser Welding

**W. W. Duley**

University of Waterloo

A WILEY - INTERSCIENCE PUBLICATION

**JOHN WILEY & SONS, INC.**

New York • Chichester • Weinheim • Brisbane • Singapore • Toronto

*Library of Congress Cataloging-in-Publication Data:*

Duley, W. W.
    Laser welding / by Walter W. Duley.
      p.   cm.
    ''A Wiley-Interscience publication.''
    Includes bibliographical references and index.
    ISBN 0-471-24679-4 (hardcover : alk. paper)
    1. Laser welding.   I. Title.
  TS228.95.D85   1998
  671.5′2—dc21                                                                    98-15242
                                                                                      CIP

Printed in the United States of America.

10 9 8 7 6 5 4 3 2

# *Preface*

Laser welding of metals is now an important industrial technology, but laser welding as a technique is still in its infancy. At first inspection, laser welding appears to be a straightforward operation; a process that can be easily implemented, and one that usually works. In reality, the joining of two metals together using laser radiation is a complex process, and involves achieving a balance between a number of competing physical and metallurgical effects. When this balance has been obtained, the result is a weld of unprecedented quality and mechanical properties. These characteristics, and the ability to weld reproducibly at high speed under CNC control with competitive cost, have made laser welding attractive in many industrial applications.

Optimization of laser welding is facilitated by the properties of lasers themselves, and the nature of laser radiation, which lends itself to tight focussing, and control over spatial and temporal profiles. Most materials that can be welded, can be welded better using laser radiation, but improvements in laser welding come only after an increased understanding of the laser welding process itself and the ways in which this process can be controlled.

The purpose of this book is to explore the inter-relationship between the process of laser welding and the results obtained. The practical goal is to produce defect-free welds, at high speed, under reproducible conditions. This requires, in the first instance, some knowledge of the properties of welding lasers and the way in which laser radiation is brought to the workpiece. Some basic information is given in chapter 2. Chapter 3 discusses the important problem of joint design and fitup in laser welding and summarizes welding data for metals under a variety of conditions, including the important new application of tailor blanking. Methods for improving laser welding efficiency in metals are found in chapter 7, while data and techniques for laser welding of ceramics, glasses, composites and polymers are given in chapter 6.

Fundamental aspects of heat transfer in laser welding, together with a discussion of the properties of the laser weldpool, are introduced in chapter 4 and are related to metallurgical properties of the laser weld in chapter 5. Results obtained on the laser welding of steels, aluminum alloys and other metals are discussed in detail in this chapter. "Laser Welding" ends with a comprehensive review of the role of diagnostics in laser welding (chapter 8), which includes signal acquisition based on plasma, optical, acoustical and vision based sensors. These are related to process monitoring and control in chapter 9. A comprehensive, but by no means complete, list of references on the subject of laser welding is also given. ''Laser Welding'' concludes with an appendix containing a specialized bibliography on specific industrial laser welding applications.

This book would not have been possible without the support of my colleagues in the laser welding field, and for their permission to quote freely from their published work. My special thanks to Debbie Guenther for word processing the final version of this book.

*W. W. Duley*

# Contents

# 1

## Introduction

### 1.1 BACKGROUND

Laser emission at optical wavelengths was first reported in 1960 with the announcement of laser oscillation in optically pumped ruby crystals (Maiman 1960). This was followed in 1961 by the first paper on laser emission from $CO_2$ gas at a wavelength of 10.6 μm (Javan, Bennett, and Herriott 1961) and then in 1964 by a report on 1.06-μm laser emission from optically pumped Nd:YAG (Geusic, Marcos, and Vankitert 1964). With the availability of these novel energy sources, many new opportunities were created for materials processing. One of these was laser welding.

By 1962, there had been several reports on metallurgical applications of lasers including welding (Bahun and Engquist 1963, Dunlap and Williams 1962, Engquist 1962). This was followed by a number of fundamental studies of laser welding (Fairbanks and Adams 1964, Anderson and Jackson 1965, Pfluger and Maas 1965, Schmidt et al. 1965, Miller and Ninnikhoven 1965, Rykalin and Kransulin 1965, Rykalin and Uglov 1965, Smith and Thompson 1967).

The advantages of laser welding in comparison to soldering of fine wires and electronic components were soon realized. Pfluger and Maas (1965) summarize these as follows:

1. elimination of unnecessary metal interfaces in the current path
2. elimination of flux
3. higher mechanical strength
4. greater resistance to vibration and shock
5. higher operating temperature potential

6. a minimum possibility of degradation of heat sensitive components during assembly

7. increased reliability

While these advantages are specific to the joining of fine wires in electronic circuits by laser welding rather than by soldering, it was apparent that laser welding could be effective in other joining applications. These were discussed in a series of early reviews on laser applications in industry (Cohen and Epperson 1968, Smith and Thompson 1967, Gagliano, Lumchey, and Watkins 1969, Gagliano and Zaleckas 1972, Ready 1971).

Emphasis was placed on pulsed laser welding as continuous-wave (cw) devices of high power were not generally available until the early 1970s. Most of these early welding studies were carried out with pulsed ruby laser radiation typically with a peak output power of 5 kW in a 1 msec pulse. Pulse energies were 1-5 J at a repetition rate of $\leq 1$ Hz. Higher pulse energies were available, but the average power of these lasers was low due to their low efficiency and attendant rod heating. Nd:YAG lasers, with their high average power capability soon offered a better alternative for spot welding and seam welding via overlapping spot welds. However, it was not until cw power levels in excess of 1 kW were available that full seam welding capability could be demonstrated.

The apparent difference between welding with laser radiation and the electron beam was noted by Ready (1971) who commented on the absence of a penetration (or keyhole) welding mode with laser radiation. With laser radiation incident on metals, the onset of keyhole welding conditions is now known to occur at an incident laser intensity of several $MW/cm^2$. In addition, the keyhole requires a finite time to become established so that it cannot be easily propagated in overlap seam welding with pulsed laser radiation.

This situation changed in 1971–72 with the announcement of welding trials using cw $CO_2$ laser radiation in the multikilowatt range (Banas 1971, Locke 1972, Locke et al. 1972, Baardsen, Schmatz, and Bisaro 1973, Locke and Hella 1974, Hoag, Pease, and Staal 1974, Ball and Banas 1974). Full penetration welds in heavy gauge stainless steel showed evidence for keyhole formation and were similar to those produced with electron beam welders (Locke and Hella 1974). These pioneering works on welding of metals with $CO_2$ laser radiation demonstrated the capabilities of high-power laser welding, but also indicated limitations. The development and optimization of high-power $CO_2$ laser welding techniques was also reported at that time by research groups in Japan (see Arata 1987 for a review), Germany (Ruffler and Gürs 1972), the United Kingdom (Swift-Hook and Gick 1973), and the Soviet Union (see the review by Rykalin, Uglov, and Kokora 1978). Subsequent advances in $CO_2$ laser welding have centered on obtaining more compact, reliable laser sources of high-beam quality and in understanding the complex interaction of joint design, welding speed, beam focussing and plasma effects in relation to weldability. With few exceptions, laser welding has not been carried out at significantly higher power than the 20 kW available in these early studies. Indeed, subsequent experience in laser welding process development has shown that there would not be much advantage to welding at laser powers in excess of 12–15 kW unless the application involves welding at very high speed or welding heavy gauge metals.

While spot welding with Nd-glass and Nd:YAG laser radiation was explored at an early stage (see the review by Gagliano and Zaleckas 1972), the capabilities of welding with 1.06-$\mu$m laser radiation were demonstrated only after average laser power exceeded several hundred watts. Because of the lower reflectivity of metals at 1.06 $\mu$m compared to reflectivity at the 10.6 $\mu$m wavelength of the $CO_2$ laser, it was found that comparable welding performance could be obtained at much lower average power. Fiber-optic delivery was also feasible at 1.06 $\mu$m even at powers exceeding 1 kW, while this capability was not available with the $CO_2$ laser. In the early 1980s Nd:YAG lasers with average output powers of 200–1000 watts were available, although high powers could only be obtained from multimode beams. Pulse energies were typically 5–100 J/pulse at repetition rates of up to 200 Hz in a 0.1–10 msec pulse.

In contrast to the situation with $CO_2$ lasers, where emphasis has been on device development without increasing the maximum output power available, the trend with Nd:YAG systems has been to higher average power. This has been limited only by the availability of techniques for the growth and excitation of high-quality crystals. Welding data has now been reported at an average power of 4 kW (Nishimi et al. 1996). The use of diode laser arrays to pump high-power Nd:YAG crystals is also an important development and will lead to more efficient laser operation with improved beam quality.

Direct diode arrays with output wavelengths in the near infrared are now becoming available with average powers extending to 1 kW and electrical-optical conversion efficiencies approaching 50%. These devices may one day challenge $CO_2$ and Nd:YAG lasers in traditional welding applications.

## 1.2 THE LASER WELDING PROCESS

Laser welding represents a delicate balance between heating and cooling within a spatially localized volume overlapping two or more solids such that a liquid pool is formed and remains stable until solidification. The objective of laser welding is to create the liquid melt pool by absorption of incident radiation, allow it to grow to the desired size, and then to propagate this melt pool through the solid interface eliminating the original seam between the components to be joined. Unsuccessful results are obtained if the melt pool is too large or too small or if significant vaporization occurs while it is present. The quality of the resulting weld may also be compromised by vaporization of alloy components, excessive thermal gradients that lead to cracking on solidification, and instabilities in the volume and geometry of the weld pool that can result in porosity and void formation.

Maintenance of the balance between heat input and heat output depends on constant absorption of laser radiation and uniform dissipation of heat inside the workpiece. The path taken by laser radiation toward the workpiece is often interrupted due to the evolution of hot gas from the laser focus. Under certain conditions this hot gas may turn into a plasma that can severely attenuate the laser beam due to absorption and scattering. Constant heat dissipation within the workpiece in the presence of a weld poor requires a stable geometry between the fusion front and the surrounding metal. The liquid-solid interface is rarely undisturbed, particularly when welding is performed on moving objects or with a moving laser beam. This introduces additional geometry-dependent terms into the cooling rate of the weld pool.

With these potential limitations it may seem surprising that welds of high quality using laser radiation are possible. However, despite the constraint that excitation and cooling be balanced in laser welding, a variety of thermal and mechanical time constants act to moderate this requirement, smoothing out fluctuations to a certain extent and allowing the establishment of stable welding conditions. A primary goal of research into laser welding has been to identify those parameters that influence the stability and reproducibility of laser welding and to develop ways to control these parameters. This begins with the laser source itself because fluctuations in output power and mode structure are converted into thermal fluctuations in the weld pool and can lead to instabilities. The highly nonlinear nature of laser material interaction processes means that certain fluctuations may grow rapidly in amplitude. This also offers the possibility of control over the laser welding process through selective modulation of the laser output.

The possibilities for precision in laser welding are enormous if the many parameters related to system excitation and response can be accurately controlled. This starts with laser radiation itself as one can see with a simple example. Consider $CO_2$ laser welding of a metal with a beam having an intensity of $10^6$ W/cm$^2$. Since the photon energy of $CO_2$ laser radiation is 0.117 eV or $0.19 \times 10^{-19}$ J the photon flux under these conditions is $5.3 \times 10^{25}$ photons/cm$^2$/sec. The statistical fluctuation in this number is $7.2 \times 10^{12}$ photons/cm$^2$/sec giving a limiting accuracy of $1.4 \times 10^{-13}$ for the photon flux. Stabilization to this degree of accuracy is unreasonable, but the small value of the limiting statistical fluctuation in photon flux at the workpiece does suggest that irradiation with laser beams whose power is kept constant to a very high degree of accuracy is possible.

Fluctuations in the response of the workpiece during welding are a dominant source of instability in the overall welding process. The bandwidth of the frequency spectrum associated with these terms extends to at least 10 MHz for plasma fluctuations. Mechanical motions, specifically those related to liquid oscillations, are characterized by a much smaller bandwidth; typically $\leq 10$ kHz. Because of these terms, laser welding produces acoustic and optical emissions. Detection of these emissions is one way the welding condition may be monitored. Recognition of components in these signals, which are diagnostic of specific fault conditions, offers the possibility of real time control to optimize welding and eliminate weld defects. This can place severe requirements on data acquisition and processing rates if signals are to be recorded, analyzed, and then fed as control signals into a feedback loop that drives a mechanical or electrical system that adjusts the system to compensate for these defects.

The two fundamental modes of laser welding are: (a) conduction welding, and

(b) keyhole or penetration welding. The basic difference between these two modes is that the surface of the weld pool remains unbroken during conduction welding and opens up to allow the laser beam to enter the melt pool in keyhole welding. These geometries and examples of weld cross sections that result from these two welding modes are shown schematically in Figure 1.1.

Conduction welding offers less perturbation to the system because laser radiation does not penetrate into the material being welded. As a result, conduction welds are less susceptible to gas entrapment during welding. With keyhole welding, intermittent closure of the keyhole can result in porosity.

Conduction and penetration modes also are possible in spot welding. The transition from the conduction mode to that in which a keyhole is formed depends on the peak laser intensity and duration of the laser pulse applied to the workpiece. Tailoring of the time dependence of the laser pulse intensity can produce a change from one type of welding mode to the other during the interaction. The weld can be initiated in the conduction mode and then converted to keyhole welding later in the interaction. It is also possible to tailor this interaction such that the keyhole, once created, can

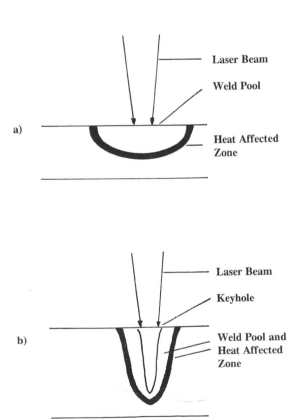

**Figure 1.1.** Comparison of conduction (upper) and penetration (lower) welding modes.

be withdrawn at the termination of the pulse in such a way that gas entrapment is minimized. The intensity temporal profile also can be adjusted to minimize the thermal gradients in the weld pool that lead to solidification cracking.

One of the primary advantages of laser welding is the ability to manipulate the laser interaction to optimize weld properties. In modern laser welding systems, this is facilitated by computer control of laser power. With feedback from an optical, plasma, or acoustic monitoring system, the laser power can often be varied in real time to compensate for changes in welding conditions. Closed-loop controllers of this type, which are becoming possible as the laser welding process becomes better understood, can ensure optimized welding under industrial conditions without operator intervention.

## 1.3  COMPARISON WITH OTHER WELDING TECHNIQUES

Laser welding is not a panacea but does have a number of advantages in comparison with conventional welding techniques. The primary advantage is the ability to narrowly focus laser radiation to a small area producing a high-intensity heat source, and then rapidly scanning this heat source along the joint to be welded. In this respect laser beam welding is comparable to electron beam welding, but has the added advantage that it can be carried out at atmospheric pressure.

With laser welding, "if you can see it you can weld it." Laser welding can be carried out in remote locations, through windows, or inside three-dimensional components where the introduction of electrodes or electron beams is impossible. Like electron beam welds, laser welding is performed from one direction only so that layered components can be welded together from a single direction. Joints that might require access from more than one direction, such as resistance spot welding of lap seams, can be laser welded from one side only. This flexibility opens up many new possibilities for joint design, particularly for components incorporating inaccessible surfaces. Some examples are discussed in chapter 3.

While laser welding systems are usually of higher capital cost than conventional welding devices, the high production rate and weld quality attainable with laser welding justifies this premium and makes such systems economically competitive. High-quality uniform laser welds can be carried out at speeds in excess of 10 m/min in mm steel stock, under CNC control, and with full diagnostic capability for weld quality. The narrow heat-affected zone of laser welds limits thermal distortion and improves metallurgical properties in comparison with arc welding techniques. The narrow heat-affected zone in laser welding together with deep penetration enhances mechanical properties including fatique strength and formability. Indeed, the entire field of tailor blanking for automotive applications, a multimillion dollar market, has been facilitated by laser welding techniques.

Laser welding also offers new opportunities in the joining of difficult materials such as aluminum and magnesium alloys and in most cases eliminates the requirement for filler material. Contamination from electrode material is eliminated with laser welding, and depletion of volatile alloy components such as zinc can be reduced.

Alloys such as the A1 7000 series that are unweldable with conventional techniques are readily joined by laser welding to produce welds that exhibit high strength and excellent formability.

There are also disadvantages of laser welding, including the high capital cost of the laser and the ancilliary systems for beam delivery and focusing. Operating costs can be high, particularly for applications that require large flows of expensive shield gases such as helium.

The tight focus attainable with laser beams, while an advantage in delivering heat efficiently to the workpiece and minimizing the heat affected zone, also brings problems of joint alignment and fit up. Small misalignments may cause large changes in welding conditions, and even narrow gaps ($\leq$0.1 mm) can result in a lack of coupling of laser radiation and reduced heating efficiency.

Laser welding of highly reflective materials such as aluminum and copper requires careful optimization of laser irradiation conditions if reflection is to be minimized. At the same time, the large thermal conductivity of these metals requires that a high laser intensity be used if the welding condition is to be initiated. This often results in the problem of reflection of laser radiation back into the laser where damage to optical components can occur. Misalignment of components during welding can also produce dangerous reflections of the laser beam.

A summary of the advantages and disadvantages of laser welding compared with other primary welding methods is given in Table 1.1. The major advantages of laser beam welding compared with other techniques derive from the fact that the heat intensity at the beam focus exceeds that attainable from conventional methods by up to several orders of magnitude (Table 1.2). This immediately introduces the concept of keyhole or penetration welding and the possibility of laser welds with a high aspect ratio, small heat-affected zone, and high processing speed.

**TABLE 1.1. Comparison of Laser Welding to Other Welding Processes**

| | | | Process | | |
|---|---|---|---|---|---|
| Parameter | LB | EB | GTA | GMA | RW |
| Joining efficiency | 0 | 0 | − | − | + |
| High aspect ratio | + | + | − | − | − |
| Small heat-affected zone | + | + | − | − | 0 |
| High processing speed | + | + | − | + | − |
| Bead profile | + | + | 0 | 0 | 0 |
| Weld at atmospheric pressure | + | − | + | + | + |
| Weld reflective metals | − | + | + | + | + |
| Combine with filler | 0 | − | + | + | − |
| Automate process | + | − | + | 0 | + |
| Capital cost | − | − | + | + | + |
| Operating cost | 0 | 0 | + | + | + |
| Reliability | + | − | + | + | + |
| Fixturing | + | − | − | − | − |

+, advantageous; −, disadvantageous; 0, neutral; LB, laser beam; EB, electron beam; GTA, gas tungsten arc; GMA, gas metal arc; RW, resistance welding

**TABLE 1.2. Intensity of Power Input at the Weld Surface for Various Welding Techniques**

| Process | Source Intensity $(W/cm^2)$ | Aspect Ratio (depth to width ratio) |
|---|---|---|
| Laser | $10^6–10^7$ | High |
| Electron beam | $10^6–10^7$ | High |
| GTA | $10^2–10^4$ | Small—medium |
| GMA | $10^2–10^4$ | Small—medium |
| Resistance | | Small |

**TABLE 1.3. Comparison Between Laser Welding and Alternative Welding Methods for Specific Joining Applications**

| Application | Comparison | Reference |
|---|---|---|
| Spot weld | Pulsed laser vs. arc | Anderson and Jackson (1965) |
| Deep section welding | Laser vs. electron beam | Laflamme and Powers (1987) |
| Pipe coils | Laser vs. TIG, MIG | Penasa, Columbo, and Giolfo (1994) |
| Automotive components | Laser vs. spot, arc | Marinoni, Maccagno, and Rabino (1989) |
| Sheet steel, coated steel | Laser vs. resistance spot, adhesive | Schmitz and Defourny (1992) |
| Steel, automotive parts | Laser vs. arc, electron beam, resistance | Iwai, Okumara, and Miyata (1987) |
| Mild steel, cold rolled, coated steel | Laser vs. MASH seam | Irving (1991) |
| Low carbon steel, tailor blanks | Laser vs. GTA | Kitani, Yasuda, and Kataoka (1995) |
| Aluminum, tailor blanks | Laser vs. GTA | Pickering et al., (1995) |
| Steel, tailor blanks | Nd:YAG laser vs. MASH seam | Sajatovic (1996) |
| Steel, tailor blanks | Laser vs. electron beam | Laflamme (1996) |
| Mild steel, deep section | Laser + arc vs. laser, arc | Matsuda et al. (1988) |
| Steel, deep section | Laser + arc vs. laser, GMAW | Magee, Merchant, and Hyatt (1990) |
| Stainless steel | Laser vs. electron beam, GTA | Heiple et al. (1983) |
| Mild steel, deep penetration | Laser vs. submerged arc | Metzbower (1983) |
| Austenitic stainless steel | Laser vs. TIG | Honeycombe and Gooch (1986) |
| High molybdenum stainless steel | Laser vs. GTA | Kujanpaa and David (1986) |
| Ti-6A1-4V | Laser vs. electron beam, GTA, plasma arc | Mazumder and Steen (1980a,b) |
| Ni alloy | Laser vs. electron beam, arc | Morochko, Fedorov, and Andree (1983) |

There have been a number of comparative studies of laser welding in relation to other welding methods in specific welding applications. A list of some of these studies is given in Table 1.3. Generally, the advantages and disadvantages noted in Table 1.1 are apparent in these comparisons, but specific applications may identify particular strengths and weaknesses. For example, laser beam welding is superior to electron beam welding of Ti-6A1-4V with regard to a reduction in porosity, but penetration is smaller (Mazumder and Steen 1980b). Weld quality and speed were comparable.

# 2

## Welding Lasers and Systems

## 2.1 INTRODUCTION

Welding of metals on a commercial basis requires laser sources of high reliability, easy operation, and low cost. In addition, the laser system must provide a typical radiative intensity of $1–5 \times 10^6$ W/cm$^2$ at the specific location on the workpiece at which welding is to be carried out. This optimized condition must be maintained during the welding operation and must be reproducible over time as a series of welds are performed. These conditions constrain not only the laser source but also the manner in which laser radiation is delivered to the workpiece. The concept of "a laser" and "a beam-delivery system" as separate entities must be abandoned in favor of an integrated approach to system design that combines all elements of laser generation and beam delivery into a single system whose performance is optimized for the specific joining application.

Virtually all laser welding systems in commercial use are based on Nd:YAG or $CO_2$ laser sources, so the present discussion is focussed on these lasers and their ancillary components. Other sources, such as the ruby laser, have been used for some welding applications, and these are discussed briefly.

New devices, such as the CO laser, are in the evaluation stage with respect to their possible use in welding applications. The most promising of new sources as a class are high-power semiconductor diode arrays. As reliability and lifetime improve and the cost per diode is reduced, diode arrays may become competitive with Nd:YAG and $CO_2$ lasers for certain welding applications. A more immediate application is as a pump source for the Nd:YAG laser; diode-pumped Nd:YAG lasers with output powers in the kilowatt range may become available soon.

## 2.2  Nd:YAG LASERS

Commercial Nd:YAG lasers for welding applications are available from many suppliers. Average output powers of 0.3–3 kW are standard, but advances in laser technology are extending the maximum power available to at least 4 kW. These lasers may be operated in three modes:

1. continuous output
2. pulsed pumping
3. Q-switched mode

Laser output characteristics in these three regimens are summarized in Table 2.1. With pulsed pumping, the pump input may be controlled to tailor the temporal shape of laser output pulses. Customized pulse shapes often are useful in optimizing welding conditions, particularly in spot welding of some A1 alloys. A complete discussion on tailoring of pulse shapes is provided by Weedon (1987).

Pulsed pumping offers a range of pulse length from ~0.1 msec to continuous-wave (CW) operation. Typical pulse durations for welding applications are 1–20 msec. At the low end of this range, pulse repetition frequencies can approach 1 kHz.

Q-switching of the laser output is less useful in welding applications because the pulse duration is much shorter ($\leq 1$ $\mu$sec), although pulse repetition frequencies can be high (up to 100 kHz). The higher peak power in these pulses facilitates plasma formation and gas breakdown.

Although penetration welding is possible in the CW mode at high average laser powers ($\gtrsim 1$ kW), this condition is not always attained at low average power. An example of this effect is shown in Figure 2.1, in which keyhole conditions are seen to be established in pulsed mode but not in CW mode at the same average power of 280 W.

The generation of high average power in Nd:YAG laser systems is accomplished by combining several individually pumped laser rods in a single resonator (Figure 2.2). This optimizes both pumping efficiency and energy extraction efficiency. The overall electrical efficiency of these lasers is low, however, with ratios of laser output power to electrical input power in the 0.5–3% range. Oscillator-amplifier

**TABLE 2.1. Output Characteristics of Nd:YAG Lasers under Different Excitation Conditions**

| Mode | Average Power (kW) | Peak Power (kW) | Pulse Duration | Pulse Repetition Rate | Energy/Pulse (J) |
|---|---|---|---|---|---|
| Continuous | 0.3–4 | — | — | — | — |
| Pulsed | To 4 | To 50 | 0.2–20 msec | 1–500 Hz | To 100 |
| Q-switch | To 4 | To 100 | <1 $\mu$sec | To 100 kHz | $10^{-3}$ |

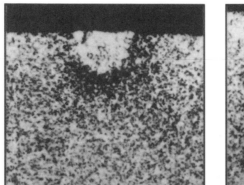
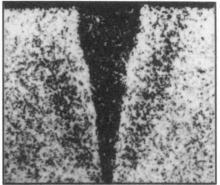

**Figure 2.1.** *Cross sections of welds (mean power [$P_m$], 280 W; welding speed [v], 2.5 mm/sec; base metal, nonalloy steel). (a) continuous mode; (b) pulsed mode. From Sepold (1984).*

configurations also may be used instead of single multirod oscillators. Pumping in these devices is customarily performed with arc lamps mounted in a close coupling optical geometry that ensures as much visible pump radiation as possible is absorbed by the laser rod. The lifetime of these lamps is typically 500–1000 hours depending on operating conditions. Replacement of arc lamps with high-power diode laser (HPDL) arrays optimizes power input to the Nd:YAG gain medium. The high electrical-to-optical conversion efficiency of HPDL diodes, which can approach 50%, will improve overall operating efficiencies for Nd:YAG systems. In addition, with less heat input, thermal distortion can be minimized, resulting in improved beam quality.

The highest laser power is obtained only under multimode conditions in which

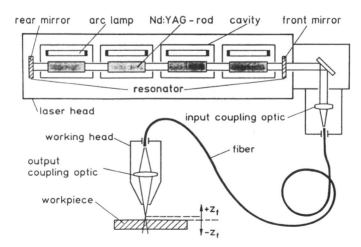

**Figure 2.2.** *Schematic of typical Nd:YAG laser system. From Tonshoff, Meyer-Kobbe, and Beske (1990).*

the filling factor within the laser cavity is large. Stable fundamental $TEM_{00}$ mode operation requires careful optimization of the optical cavity and minimization of thermal distortion within the laser rods. In general, this constrains operation to low pumping rates and low cavity filling factors. Output power in the fundamental mode then is substantially less than that obtained during multimode operation. A complete discussion on this subject is provided by Koechner (1992).

One of the prime advantages of the Nd:YAG laser over the $CO_2$ laser is the ability to deliver laser radiation through optical fibers. Fortuitously, the 1.06-$\mu$m output wavelength of the Nd:YAG laser falls within the wavelength range in which glass fibers have low attenuation, so propagation of Nd:YAG laser radiation over distances of as much as several hundred meters is possible with minimal loss. This is attractive in robotic or multiaxis laser welding applications. A quantitative study of the power-dependent transmission loss in long optical fibers was conducted by Ishide et al (1990). An example for a 200-m-long graded index fiber is given in Figure 2.3. Other studies of fiber transmission of high-power Nd:YAG laser radiation were conducted by Nakajima et al. (1989) and Miura and Shibano (1990). Optical design considerations for fiber input and output couplers are discussed by Gascoin et al. (1987), Notenboom, Nonhoff, and Schildbach et al. (1989), Bloehs and Dausinger (1995), Keicher and Essien (1995), and Hunter et al. (1996). Near-diffraction limited performance with good standoff distance is possible in optimized systems. A detailed

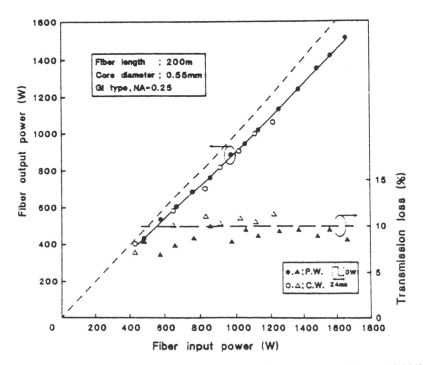

**Figure 2.3.** *Transmission loss of 200-m optical fiber (GI type). From Ishide et al. (1990).*

study of the effect of focus on penetration and bead width in welding with fiber-optic transmission was reported by Miura and Shibano (1990) and Tonshoff, Meyer-Kobbe, and Beske (1990).

With fiber-optic coupling, it is possible to combine the power from two or more lasers. One such combination is shown in Figure 2.4, the output from three 1-kW lasers is combined into a single beam. The transmission loss in this system was reported to be 24.7% (Norris et al. 1992). An alternative approach is to time share the output from a single laser among several locations by beam switching into different fibers.

An interesting application of beam combination in laser welding was reported by Narikiyo et al. (1995); the beams from two 2-kW and one 1-kW Nd:YAG laser were combined at the weld focus (Figure 2.5). The 1-kW laser was pulsed, whereas the two 2-kW lasers were operated in CW mode. The effect of different orientations of these three beams on penetration depth versus welding speed is shown in the figure. Each beam was found to create its own keyhole at the laser focus.

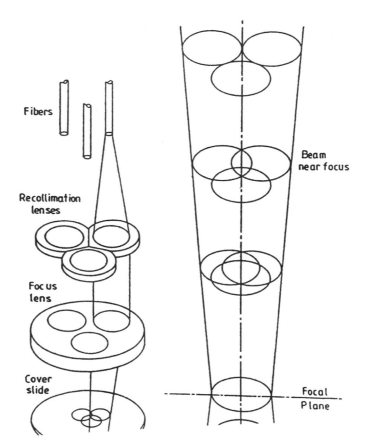

**Figure 2.4.** Schematic of Multilase focussing assembly. From Norris et al. (1992).

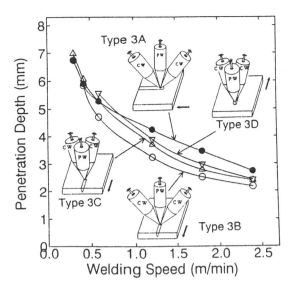

**Figure 2.5.** *Relationship between penetration depth and welding speed in various arrangements of focussing optics and welding direction (CW 2 kW + pulsed wave 1 kW + CW 2 kW). From Narikiyo et al. (1995).*

Some additional reports on Nd:YAG laser systems for the welding of metals are given by Aubert et al. (1987), Seiler (1988), Manes and Zapata (1990), Jellison, Keicher, and Fuerschbach (1990), Huang et al. (1993), and Haruta (1995).

## 2.3 CO₂ LASERS

The $CO_2$ laser remains the workhorse of industrial laser welding systems because it is simple and reliable and is available as a robust well-engineered industrial machine with output powers of up to 50 kW. The trend in laser welding applications has been toward higher powers, with CW $CO_2$ lasers of 5–10 kW routinely used in production welding of a variety of components.

The $CO_2$ laser is an efficient device, at least compared with other lasers, such as the Nd:YAG. Output efficiency, defined as the ratio of output laser power to input electrical power, can approach 10%. The laser mixture contains $N_2$ and He gas in addition to $CO_2$ and flows in a closed system in which most of the gas is recycled through the discharge. This lowers operating costs.

The laser discharge can be run as a true direct current (DC) discharge for CW output and electrically pulsed at kilohertz frequencies to produce output pulses as short as 0.1 msec. The peak power in such short pulses can be an order of magnitude greater than the average power at repetition rates of a few kilohertz. Although a DC glow discharge is easily sustained in flowing $CO_2$ lasers, the instabilities inherent in this mode of excitation have led to the development of high-frequency– (~150 kHz) and radiofrequency (RF) Excited (13.5 and 27 MHz) devices (Hishii, Sato, and Fukushima 1987, Yagi, Kuzumoto, and Ohtami 1992, Wollermann-Windgasse et al. 1986).

Capacitive coupling in these lasers eliminates electrodes within the gas and lowers the excitation voltage applied to the discharge. The full volume of the gas within the laser resonator also can be excited, leading to high gain and output power per unit length of discharge. The absence of electrodes minimizes gas contamination and gas consumption. High-frequency or RF excitation also facilitates pulsing of the discharge with pulse repetition frequencies of 100 kHz in RF-excited devices and pulse lengths as short as 10 $\mu$sec.

The beam output can be constrained to a $TEM_{00}$ or gaussian mode, often with little reduction in overall output power in commercial lasers. Low-order multimode output ($TEM_{01*}$,) is used in many welding applications. A typical beam profile is shown in Figure 2.6. Unstable resonator configurations also can be used when an annular mode is required; this is more useful for heat treating than for welding applications. In an annular beam, the peak intensity is reduced from that of a uniform intensity beam of the same outer diameter by $(M^2 - 1)/M^2$, where M is the magnification factor. An M value of 2–4 is common in unstable resonator configurations.

Output beam diameter scales with laser power but typically is 20–30 mm to the $1/e^2$ points in a Gaussian or quasi-Gaussian output beam. Beam divergence is $\leq 1.5$ mrad in $TEM_{00}$ or mixed low-order mode operation. A summary of representative technical data for $CO_2$ lasers used in welding applications is given in Table 2.2.

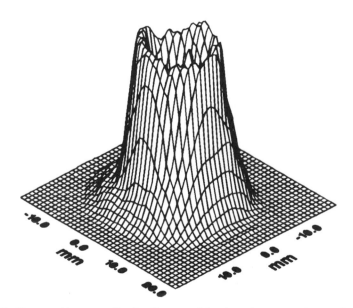

**Figure 2.6.** *Measured beam profile at a distance of 5 m from the output of a 10-kW RF-excited $CO_2$ laser. From Hertzler and Wollermann-Windgasse (1994).*

TABLE 2.2. Representative Data for Commercial CO₂ Lasers Used in Welding Applications

| | Output Power (W) | | | | | | | | | | |
|---|---|---|---|---|---|---|---|---|---|---|---|
| | 1100 | 1750 | 2200 | 3000 | 3000 | 6000 | 6000 | 8000 | 10,000 | 25,000 | 45,000 |
| Power range (W) | 200–1200 | 200–1800 | 110–2200 | 150–3000 | 500–3150 | | 300–6000 | 400–8000 | 600–12,000 | | |
| Mode | $TEM_{00}$ | $TEM_{00}$ | $TEM_{00}$ | $TEM_{00}$, $TEM_{01^*}$ | | Unstable resonator | $TEM_{01^*}$ | $TEM_{01^*}$ | Low order | Unstable resonator | Unstable resonator |
| Beam diameter (mm) | 15.5 | 17.7 | 13 | 20 | 15.5 | 50 | 22 | 28 | 31 | 50 | 65 |
| Beam divergence (mrad) | | | <1.5 | <1.5 | | | <1.5 | <1.5 | <1.5 | | |
| Pulse width | 0.1–10 msec | 0.1–10 msec | 10 μsec –CW | 10 μsec –CW | 0.1–10 msec | | 10 μsec –CW | 10 μsec –CW | 10 μsec –CW | | |
| Pulse rate | 10–1000 Hz | 10–1000 Hz | 100 Hz–100 kHz | 100 Hz–100 kHz | 10–2000 Hz | | 100 Hz–100 kHz | 100 Hz–100 kHz | 100 Hz–100 kHz | | |
| Gas consumption (l/hour) | Total, 40 | Total, 50 | | | Total, 73.4 | | | | | | |
| CO₂ | | | 1.5 | 1.5 | | 12 | 1.5 | 2.0 | 2.0 | 24 | 24 |
| N₂ | | | 6.0 | 6.5 | | 100 | 6.5 | 10 | 10 | 200 | 200 |
| He | | | 18.0 | 32.0 | | 180 | 32 | 48 | 48 | 360 | 360 |

17

Various resonator and gas flow configurations are used in these devices, which accounts for the variation in output parameters and gas consumption. A discussion of various gas flow, excitation, and resonator configurations used in high-power $CO_2$ laser devices is provided by Duley (1976), Wollermann-Windgasse (1986), Hishii et al. (1987), Loosen (1992), Macken (1992), Hügel (1987), Sugawara et al. (1987), Hertzler and Wollermann-Windgasse (1994), and Rath and Northemann (1994).

Although carefully selected transmissive optics such as ZnSe can be used as output windows on high-power lasers, thermal distortion can be a problem. This has led to the development of the aerodynamic window (Loosen 1992), in which a rapid transverse flow of gas at high pressure is directed across the output aperture (Figure 2.7). This gas flow is ~5 $m^3$/min at an initial pressure of 5 atm.

A diagnosis of the laser beam profile is difficult at high laser power, but a variety of commercial scanning wire or scanning aperture systems are available that can be placed in the laser beam. These devices sample the beam cross section by momentarily intercepting a fraction of the beam and directing this part of the beam to a detector. With a programmed scan through the beam, the profile can be synthesized and displayed on a video monitor. Figure 2.6 shows an example of such a plot. The basic construction of one type of laser beam analyzer is shown in Figure 2.8. Other types of laser beam analyzer devices are described in Gregersen and Olsen (1990), Sasnett and Hurley (1994), Lim and Steen (1984), and Austin (1986). The use of burns in acrylic to image $CO_2$ laser beam profiles was discussed by Whitehouse and Nilsen (1990) and Miyamoto, Maruo, and Arata (1984). Imaging of beam profiles through ablation of acrylic material is useful for a preliminary estimate of beam properties but in general cannot be relied on to provide quantitative results.

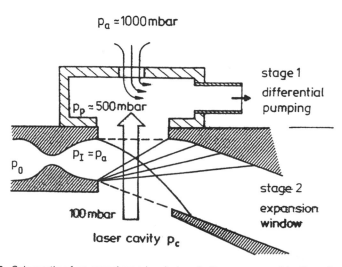

**Figure 2.7.** Schematic of an aerodynamic window, built up as a combination of an expansion window and a differential pumping stage. From Loosen (1992).

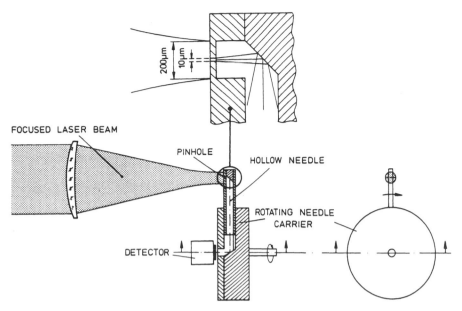

**Figure 2.8.** *Laser beam analysis diagnostic system with hollow needle and pinhole. Pinhole typically has a diameter of 10 μm. Rotation of the needle and shifting the carrier provide complete information on the intensity distribution with a resolution of 10 μm. From Beyer et al. (1986).*

Examples of the way in which laser beam analysis measurements may be used to optimize optical design in $CO_2$ laser systems are given by Essien and Fuerschbach (1996) and Beyer et al. (1986).

The direction of the $CO_2$ laser beam from the laser to the workpiece inevitably involves one or more reflections followed by focussing. Optimized performance in such a beam delivery system requires minimization of optical aberrations, the elimination of thermal distortion in optical components, and no opportunity for phase distortion as the laser beam propagates through the intervening atmosphere from the laser to the workpiece. Although the latter effect can be eliminated by enclosing the laser beam and flushing the enclosure with dry nitrogen, constraints on optical components can be severe, especially considering the high incident intensities that are involved.

The general criterion for optical finish and thermal distortion in reflective optics is λ/20 at the $CO_2$ laser wavelength. This amounts to ~0.5 μm and requires diamond-turned optics and efficient cooling. The absorbance of mirrors in such high-power applications is typically <0.5% at a 45° angle of incidence but may be polarization dependent.

A number of focussing arrangements can be used with high-power $CO_2$ laser beams (Figure 2.9) and reflective optics. The simplest involves a single parabolic mirror and results in minimal aberration despite its off-axis geometry. The focussing characteristics of 90° off-axis paraboidal mirrors in this application were investigated by Kalberer et al. (1994).

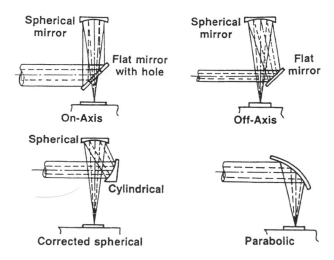

**Figure 2.9.** *Multikilowatt laser beam focussing heads. From Duhamel (1988).*

The need to accommodate motion of the optical system along the focal axis can be met by incorporating one or more extra reflective elements in the focussing head. An example of such a configuration that also uses adaptive optics to widen the range of control over focal parameters is shown in Figure 2.10. The adaptive optical components are diamond-turned copper mirrors whose curvature is dynamically controlled through changes in the pressure of liquid coolant (Figure 2.11). The performance of this system is discussed by Bea, Giesen, and Hügel (1993). The beam diameter at the focus could be changed over a range of a factor of ~5, whereas focal position itself could be moved by up to 30 mm in a 408-mm focussing system.

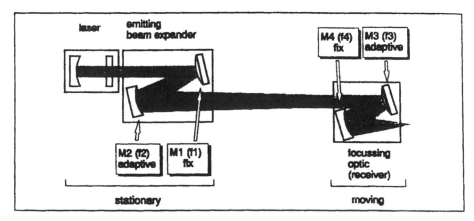

**Figure 2.10.** *Schematic of the adaptive focus system consisting of two flexible beam expanders (transmitter and receiver). From Bea, Giesen, and Hügel (1993).*

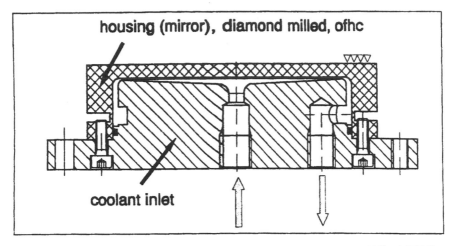

**Figure 2.11.** Schematic of the adaptive mirrors. From Bea, Giesen, and Hügel (1993).

A discussion of the relative merits of various materials used for $CO_2$ laser optical components is given by (19), McIver and Guenther (1986), Sherman et al. (1987), and Miyata (1986). The thermal response and distortion of optical components in high-power $CO_2$ laser systems are discussed by Berger (1986), Borik and Giesen (1988), Miyamoto, Nanba, and Maruo (1990), Brandon (1992), and Giesen (1992). Polarization characteristics of reflective optics for $CO_2$ lasers were studied by Iwamoto, Ebata, and Namba (1992).

IR waveguides for transmission at 10.6 μm of powers up to ~1 kW were demonstrated under laboratory conditions (Engel and Fontaine 1989, Hongo et al. 1991, Matsumoto, Seki, and Yasuda 1992, Morrow and Gu 1993, and Gu 1995). These guides are based on hollow cylindrical tubes with an internal surface coated to yield enhanced reflectivity at 10.6 μm. Typical inside diameters are 1–2 mm and are formed from metallic tubes or coated dielectric materials. Losses over 1–1.5-m lengths can be as small as 5% at incident powers of 1–2 kW (Matsumoto et al. 1992), but this is highly dependent on bend radius.

## 2.4    THE Nd:YAG LASER VERSUS THE CO₂ LASER

Until recently, high laser powers (>3 kW CW) could be obtained only with $CO_2$ lasers. As a result, $CO_2$ lasers dominated the market for industrial applications requiring powers in this range. With the rapid development of Nd:YAG laser technology, YAG lasers with CW powers up to 4 kW became available (Nishimi et al. 1996), so $CO_2$ and YAG laser systems can be directly compared in many practical applications. The question of which laser (or laser system) is best for specific applications then arises and has led to some debate.

In laser welding, the clear advantages of the Nd:YAG laser are:

1. enhanced coupling to reflective metals
2. increased processing efficiency compared with $CO_2$ lasers with the same power
3. fiber-optic delivery

The $CO_2$ laser has other advantages:

1. high electrical efficiency
2. low operating cost
3. easily scaled to high powers (e.g., 45 kW)

Representative welding data for mild steel (Figure 2.12) obtained at the same laser power (3 kW) show that higher penetration is obtained with YAG welding than with $CO_2$ laser radiation, except for thick material, for which comparable results are obtained. For thicknesses near 2 mm, YAG welding is ~50% faster than $CO_2$ laser welding. Weld quality for mild steel is similar for the two systems, an increase in welding speed may translate into cost savings in high-volume applications such as laser welding of tailor blanks. A comparison between Nd:YAG and $CO_2$ laser

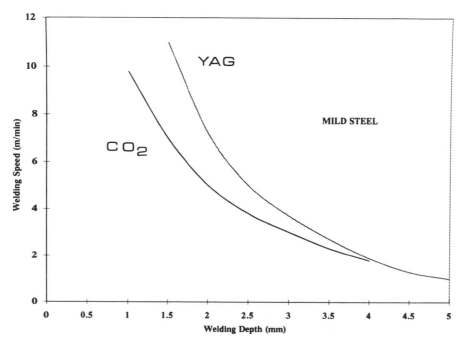

**Figure 2.12.** Comparison of welding speed versus weld depth for 3-kW YAG (data from Haas Laser, Schramberg, Germany) and 3-kW $CO_2$ (data from Trumpf Inc., Ditzingen, Germany) Laser welding of mild steel.

radiation in welding of Al 6082 alloy (Dausinger et al. 1996) shows that the intensity threshold for keyhole initiation with 1.06-μm radiation is almost half that required at 10.6 μm. However, process limits appear to be more tightly defined at 1.06 μm than at longer wavelengths.

The primary advantages of Nd:YAG compared with $CO_2$ lasers are reduced primarily to the flexibility offered by fiber-optic delivery. This is particularly true in multiaxis gantry or robotic beam delivery systems. (Iwai, Okumura, Miyata 1987)

## 2.5  BEAM DELIVERY

Presentation of the laser beam to the workpiece is critical to successful laser welding. The parameters to be controlled include:

1. position of beam centerline relative to weld seam
2. position of beam focal plane relative to surface
3. beam intensity (spatial and temporal distribution)
4. scan speed
5. ambient atmosphere at weld site, along trailing edge, and on underside of weld

In practical welding applications, these factors are also influenced by part geometry and the necessity to maintain fitup to high tolerance over the entire weld operation. This in turn places severe constraints on clamping of parts, edge preparation before clamping, and system design to minimize distortion.

For certain applications, such as one-dimensional (1D) welding of tailor blanks, an acceptable solution can be obtained with a relatively simple single-axis welding station in which parts are moved along a preset linear track below a fixed welding head. Alternatively, the laser beam can be scanned along the joint between two fixed parts. In either of these geometries, the laser focus is set at the appropriate location relative to the surface of the weld and is not adjusted during welding. The position of the laser beam centerline also is placed at a fixed location relative to the joint between the components and relies on edge preparation and sound mechanical clamping to maintain alignment during welding. Scan speed and laser intensity may be controlled during welding to optimize welding quality.

Two-dimensional welds in the blank configuration can be obtained through the addition of a second axis to the previous 1D system. This axis can be incorporated into the beam scanning unit or used as a separate motion of the welding table below a fixed laser beam. A hybrid configuration in which one axis is assigned to beam motion while the other is contained within a motion table for the part also is possible. Two-dimensional welds also are suited for robotic beam delivery.

The welding of three-dimensional (3D) parts can be accomplished with gantry or robotic beam delivery systems. In principle, a robot also may be used to manipulate a part under a fixed laser beam to effect a 3D weld, but joint fitup and clamping often mitigate against such a system. With gantry and robotic delivery, the laser beam is positioned using a combination of three translational and two rotational axes in a five-axis system. Both Nd:YAG and $CO_2$ lasers can be used in such five-axis systems, but the availability of fiber-optic beam delivery favors Nd:YAG lasers in this application.

Position sensing through the use of optical emission or another feedback mechanism is important in the welding of 3D parts because part flatness may be inconsistent or distortion during welding can introduce dimensional irregularities.

In 3D welding, the weld track must be created as a 3D contour that relates position in the workpiece to that in "world" coordinates. This contour must be compatible with the kinematic capabilities of the motion control system. The philosophy and practicalities of such requirements in specific welding applications were discussed by Sepold and Zierau (1992), Steen and Li (1988), Schraft et al. (1988), Charles et al. (1988), Bea et al. (1994), and Matsumoto et al. (1996), Hügel et al. (1995). Sensor outputs can be used to detect path location for comparison with predicted location, and generation of an error signal or active sensing can be implemented in a seam detection mode. Because positional accuracies of 0.1 mm must be implemented to ensure optimized welding conditions, stringent requirements are placed on both data acquisition and processing rates, as well as on the feedback and control systems. Although welding speeds in excess of 10 m/min are feasible in one- and two-dimensional geometries with feedback control over path trajectory and beam focus, they often are not practical in the welding of 3D parts, when rapid changes in path contour are encountered.

# 3

# *Laser Welding of Metals*

## 3.1 INTRODUCTION

The welding of metals was one of the first industrial applications of lasers. The ability to direct a pulse of intense radiation to a remote location using only an optical system was soon realized to be enabling technology. The result has been the rapid development of laser welding techniques and the adoption of laser welding as a standard process for joining metals under a variety of industrial conditions.

This chapter reviews the laser weldability of metals and outlines the laser welding process as a whole with emphasis on the specification of optimized welding parameters. A large database has been accummulated on laser welding of metals, and this is summarized, although a complete survey has not been attempted.

The important concerns of joint design and the mechanical properties of laser-welded seams also are discussed. The developing process of laser welding in the manufacture of tailor blanks for automotive structures is reviewed separately.

## 3.2 WELDABILITY

The weldability of two metals relates not only to alloy type and composition but also to the practicality of welding these particular materials together under the conditions required. Thus, the weldability that may be acceptable under laboratory or prototyping conditions may be unacceptable as a production process because of such considerations as limits to welding speed, requirements for joint fit up, preheating, and so on. Such constraints must be superimposed on metallurgical properties such as the tendency for cracking, brittleness, and pore formation.

Although laser welding offers new flexibility in the joining of metals and laser welds are usually of high quality, laser welding is not a panacea and high-quality laser welds are obtained only after optimization of key process variables. Some of these variables include:

- joint design and preparation
- weld thermal cycle
- gas flow and composition
- preheating (if necessary)
- filler type and feed rate
- changes in alloy composition
- thermal effects in heat-affected zone

Some generalizations on the weldability of various metals are summarized in Table 3.1, and information on the weldability of dissimilar metals is given in Figure 3.1. In the welding of dissimilar metals, good solid solubility is essential for sound weld properties. This is achieved only with materials with compatible temperature ranges for melting, such as Ni and Co. In other materials, such as Al and Fe, for which the melting temperature of one component is near the vaporization temperature of the other, poor weldability is obtained and often involves the formation of brittle

**TABLE 3.1. Weldability of Various Metals**

| | |
|---|---|
| Low-carbon and High strength low alloy steels | Good welds with $CO_2$ and Nd:YAG lasers; high formability |
| Medium- and high-carbon steels | Acceptable but may be subject to cracking; brittle, low ductility |
| Alloy steels (e.g., pipeline, ship building, structural) | Can be welded satisfactorily but root porosity in partial penetration welds; high weld hardness |
| Galvanized, galvanneal steels | Good welds and high speed but joint design important in lap weld |
| Austentitic stainless steel | Excellent welds, low porosity, good corrosion resistance |
| Ferritic stainless steels | Acceptable but welding reduces toughness and corrosion resistance; HAZ may have high hardness |
| Martensitic stainless steels | Hard, brittle welds; needs preheating or postweld tempering |
| Heat-resistant alloys (e.g., Hastelloy, Inconel, INCO 718) | Evaluate case by case; brittleness, segregation, cracking may be problem; filler may be useful |
| Titanium and Ti alloys | Good welds in Ti-6A1-4V and Ti; inert gas shielding necessary, good mechanical properties |
| Aluminum and A1 alloys | Good welds possible with $CO_2$ and Nd:YAG lasers under controlled conditions; loss of volatile elements leads to porosity; cracking a problem in 6000 series |
| Magnesium 1 | Good $CO_2$ laser welds in Mg Al 9 Zn |
| Copper | Acceptable welds possible in thin sheet with Nd:YAG radiation; can be spot welded |
| Brass | Difficult due to loss of Zn and high thermal conductivity |

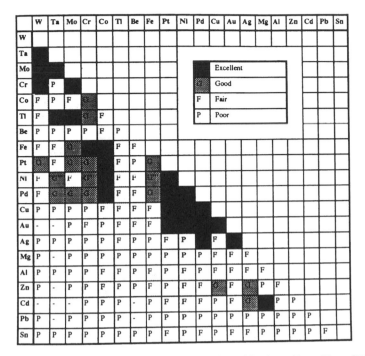

**Figure 3.1.** *Laser weldability of dissimilar metal combinations. From Steen (1991).*

intermetallic phases. A recent review of laser welding of dissimilar metal combinations can be found in Sun and Ion (1995).

### 3.2.1  Welding Parameters

In addition to laser type, laser power and weld speed, which determine the rate of energy input to the workpiece, successful laser welding requires optimization of additional parameters such as the size and location of the focal spot and the type and flow rate of shielding gas and nozzle geometry. A checklist is given in Table 3.2.

Optimization of these parameters defines the welding condition and can be done only in relation to what is a "good" as opposed to a "bad" weld. In many cases, this determination involves sectioning welds to measure weld cross section and shape in addition to metallurgical and microstructural properties. Mechanical testing also will define a range of acceptable welding conditions.

### 3.2.2  Laser Power

The operational range of a laser welding system is the first parameter defined through the relation between laser power and welding speed for a given material

**TABLE 3.2. Critical Parameters in Laser Welding**

Laser power
Amplitude
CW or pulsed wave
CW plus pulse
Pulse shape and repetition rate
Focussing
Location of focal spot on surface
Focus above or below surface?
Depth of focus
Intensity distribution within spot on surface
Shield gas
Type
Flow rate
Orientation relative to focus
Trailing, coaxial, leading?
Reverse side
Nozzle flow pattern

and specified weld penetration depth. This curve, or set of curves, is derived empirically on the system after some initial guidance from previous experimental data. An example of such a relation for $CO_2$ laser welding of mild steel is shown in Figure 3.2. The curves in this figure specify a range of conditions that generally lead to acceptable welds, although optimization based on criteria other than penetration depth may narrow these conditions to a more restricted range.

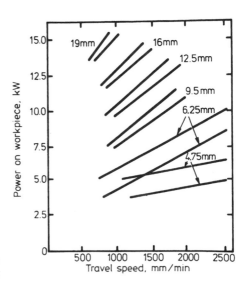

**Figure 3.2.** $CO_2$ laser power as a function of welding speed for different thicknesses of mild steel plates. From Metzbower (1983). (Reprinted with permission. Copyright TWI)

Laser power as a function of welding speed for different thickness mild steel plates.

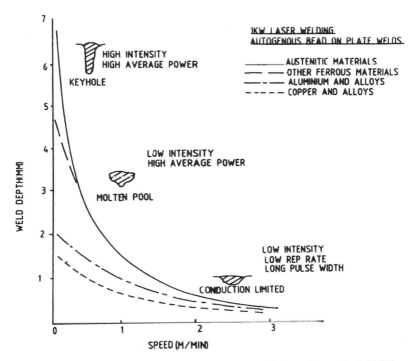

**Figure 3.3.** *Welding performance of 1-kW Nd:YAG laser. From Hoult (1990).*

Data may also be presented as weld depth, d, versus weld speed, v, at constant laser power. Figure 3.3 summarizes the capability of a 1-kW average power Nd: YAG laser in welding of different metals and shows the shape of the weld profile under different welding conditions. The functional form of the penetration depth or weld speed curve in this figure is commonly observed in laser welding and approximates d = c/v, where c is a constant, at least over part of the welding range.

Bead width also can be measured and correlated with welding speed. The shape of a typical weld profile is shown in Figure 3.4. In general, the width of the bead

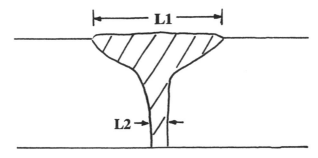

**Figure 3.4.** *Weld bead configuration.*

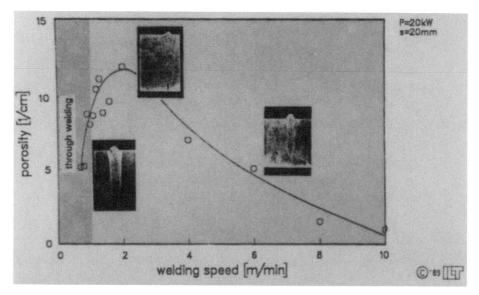

**Figure 3.5.** *Porosity of low-carbon steel of 20-mm thickness versus welding speed at 20-kW beam power. Shaded area indicates the region of deep-penetration welding. From Funk et al. (1989).*

at the top of the weld (L1) is greater than that deeper into the weld (L2), although the difference between these two values depends on weld speed and incident laser intensity. Kimara, Sugiyama, and Mizutame (1987) reported that the difference between L1 and L2 decreases with welding speed at constant laser power.

Porosity is another parameter that can be significant in assessment of weld quality and its dependence on processing conditions. Porosity is especially important in deep welds without full penetration, as shown, for example, in steel (Figure 3.5). It also is important in the welding Al alloys containing volatile constituents (Figure 3.6). The correct set of operating conditions may minimize this problem. It was shown by Richter, Eberle, and Maucher (1993) that a combination of pulsed-wave and CW excitation is effective in reducing porosity.

### 3.2.3   Focussing

The determination of optimum focussing conditions was discussed in some detail by Dawes (1992). One might expect focussing the laser beam at the surface of the weld piece to be sufficient to ensure good results, but this often is not the case. The first problem is in estimating where the focus actually occurs; nearly the same result often is obtained with test burns at locations near the focus. This is especially true when the focussing system has a large f/number and therefore a large depth of focus. The use of a wedge-shaped acrylic sample, which is scanned through the beam, is one method of locating the focal plane. Another method involves calculation of the position of the focal plane from analysis of the beam profile on both sides of the focus.

Although the laser intensity is highest in the focal plane, it decreases along the optic or laser beam axis in either direction away from the focal plane as the laser beam diverges. Creation of an extended structure, such as the keyhole required for penetration welding, often is not facilitated by having the laser focus at the front surface of the workpiece. Instead, such conditions can be created through focussing at a point somewhere inside the surface. An example of this effect from the experimental study of Matsumura et al. (1992) on $CO_2$ laser welding of Al alloys is shown in Figure 3.7. In this figure, negative focal point refers to distances below the surface of the workpiece. The increase in penetration depth seen on focussing inside the surface arises in part from the enhancement in multiple reflections.

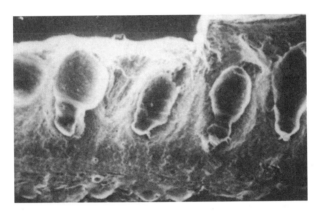

$$P_O = 3 \text{ kW}, \quad v = 2 \text{ m/min}, \quad f_d = -1 \text{ mm}$$

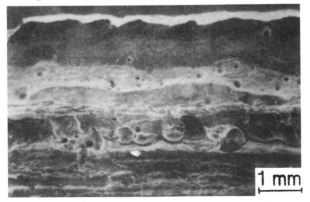

$$P_O = 3.5 \text{ kW}, \quad v = 5 \text{ m/min}, \quad f_d = -4 \text{ mm}$$

**Figure 3.6.** *Scanning electron microphotographs of two fractured weld fusion zones of laser-welded Al 2090 alloy showing porosity formation tendency depending on welding conditions. From Katayana et al. (1989). (Reprinted courtesy A&M International)*

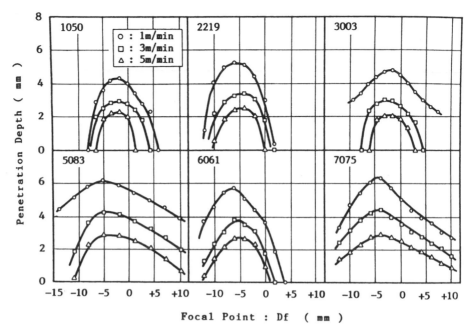

**Figure 3.7.** *Influence of focal point on penetration depth at various welding speeds in Al alloys. From Matsumura et al. (1992).*

The criticalness of focussing increases dramatically as the f/number is reduced. This arises as the depth of focus decreases with f/number, so small changes in position relative to the focal plane result in a large change in intensity. Where possible, the largest f/number consistent with the intensity required for laser welding at the chosen speed should be used. With $CO_2$ lasers, this is typically f/6–f/7, whereas with Nd:YAG laser radiation, a value near f/4 is commonly used.

The position of the focal spot relative to the joint to be welded also is an important consideration and often leads to quite stringent requirements regarding joint fit up, gap separation, and stability in the motion control system. These considerations are discussed in detail in later sections. During welding of components with different thicknesses, metallurgical properties, or both, it often is advantageous to offset the beam profile to either side of the seam. For example, when butt welding heavy-gauge material to lighter-gauge material, it may be advantageous to offset the center of the beam focus to allow more laser power to impinge on the thicker material. This ensures full penetration on both sides of the weld without the possibility of burn through the thinner component.

Under certain conditions, it has been shown (see Chapter 7) that separation of the laser beam into two components with different intensity distributions and focal characteristics can lead to improvements in welding capability.

### 3.2.4  Shield Gas

A gas flow coaxial with the laser beam or impinging on the laser focus from the side serves several purposes and is a common feature in laser welding systems. The first role of this gas often is to prevent oxidation of the weldment and creation of slag in the vicinity of the fusion zone. A second, and critical, function is to suppress plasma formation in the vapor over the weld zone and to blow away any plasma that may be created in the welding process. The latter role ensures that the laser beam can reach the weld zone with minimal interruption, thus improving weld quality and enhancing the uniformity of such factors as penetration depth and weld bead profile. It is not unusual to have both coaxial and sideflow gas nozzles directing gas at the weld region.

Helium and Ar are commonly used as shield gases, with He the gas of choice because of its high ionization potential and resistance to breakdown. Nitrogen also is commonly used as a replacement for He and has many of the same properties but is much cheaper. Gas flow rates are typically 10–40 $\ell$/min, with the higher rates necessary for high-speed welding ($>10$ m/min). The effect of optimization of gas flow on weld penetration can be seen in Figure 3.8, in which the use of $N_2$ gas rather than He leads to plasma ignition and a limited welding range (Behler et al.

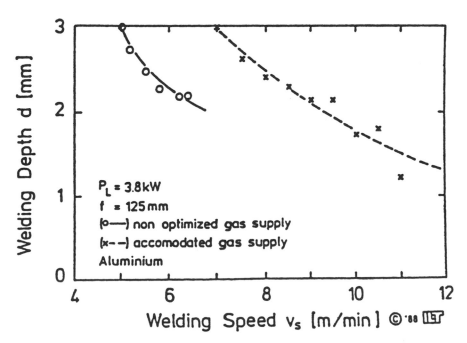

**Figure 3.8.** *Plasma ignition and the welding result are influenced by the type of assist gas. With the parameters shown and similar welding speeds, no plasma ignition is observed with He. However, with $N_2$ as an assist gas, plasma is generated. From Behler et al. (1988).*

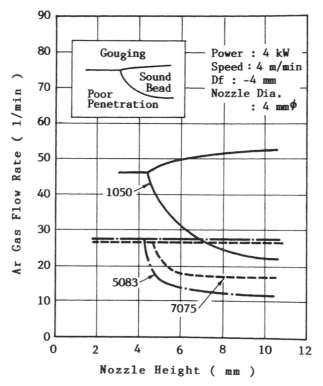

**Figure 3.9.** *Influence of assist gas and focal position on weld bead formation in CO₂ laser welding of various Al alloys. From Matsumura et al. (1992).*

1988b). Douay et al. (1996) reported that under certain conditions, the plasma may be completely suppressed by a transverse flow of He gas.

Gas flow through the weld region also can act to destabilize liquid flow, leading to poor-quality welds (Figure 3.9). These flow effects are a strong function of nozzle geometry, and a variety of nozzle designs have been developed to minimize this problem (Dawes 1992, Fieret, Terry, and Ward 1986, Steen 1991, Faerber 1997). Under certain conditions of gas flow from a coaxial nozzle, a *negative* pressure may be exerted along the beam axis, drawing material back out of the laser weldment.

Back shielding of the root of the weld is necessary when welding oxidizable metals such as Al and Ti. Argon or $N_2$ often is acceptable in this application, although He gas is preferred because it rises to the surface of the workpiece when introduced below the weld.

Excellent welding results also can be obtained at reduced pressure, at which the lower pressure prevents the formation of a shielding plasma. Deep-section $CO_2$ laser welding of heavy-gauge materials (Arata 1987) is possible in a vacuum, but there seems to be little advantage in reducing the ambient gas pressure below ~0.1 Torr because lower pressures do not lead to corresponding increases in penetration depth.

Under certain conditions, $CO_2$ laser radiation can propagate through liquid water for distances of 1–2 mm by creating its own waveguide-like structure (Dunn, Bridger, and Duley 1993). The laser intensity is attenuated in this process, but Shannon et al. (1994a,b) demonstrated that acceptable welds can be produced in underwater laser welding of steel. Weld properties were found to be similar to those carried out in air when an account was taken of beam attenuation.

## 3.3 LASER WELDING DATA

Extensive empirical data exist regarding the relationship among laser power (P), welding speed (v), penetration depth (d), bead width (W), and weld properties for many metals. These data are presented in many cases as plots of measurable quantities, such as penetration depth or bead width versus laser power and welding speed, but in general these plots have a number of common characteristics. These are:

1. an increase in laser power at constant welding speed yields a proportional increase in penetration depth,
2. an increase in welding speed at constant laser power results in a corresponding decrease in penetration depth,
3. for a given penetration depth, the welding speed scales with laser power.

The analytical fit to these data can be described on average by the simple regression equation

$$\frac{P}{vd} = a + \frac{b}{v}. \tag{3.1}$$

Values of $a$, $b$, and the regression coefficient $r$ for fits to experimental laser welding data are summarized in Table 3.3.

It is apparent that a relatively simple correlation must exist among $P$, $v$, and $d$ to support these relationships. The form of such a correlation has been investigated

**TABLE 3.3. Fit Parameters a and b (see Equation 1) and Regression Coefficient, r, for Laser Welding Data**

| Metal | Laser | a (kJ/mm$^2$) | b (kW/mm) | r |
|-------|-------|-----------------|-------------|------|
| 304 Stainless Steel | $CO_2$ | 0.0194 | 0.356 | 0.82 |
| Mild steel | $CO_2$ | 0.016 | 0.219 | 0.81 |
|  | Nd:YAG | 0.009 | 0.309 | 0.92 |
| Aluminum alloys | $CO_2$ | 0.0219 | 0.381 | 0.73 |
|  | Nd:YAG | 0.0065 | 0.536 | 0.99 |

Units are P (kW), v (mm/second), d (mm).

by many groups, such as Duhamel and Banas (1983), Mannik and Brown (1990), Metzbower (1993a,b), and Fuerschbach (1994,1996).

Although excellent parametric fits can be obtained when a full experimental data set is available (e.g., Mannik and Brown 1990), extrapolation of these parametric fits to other experimental conditions may lead to inaccurate predictions. This lack of reproducibility often can be attributed to such factors as laser mode, focal characteristics, and shielding gas flow. Weld quality, or specifically what constitutes an acceptable weld under given conditions, also must be considered, but it involves another set of variables that may be difficult to specify quantitatively. For example, at high welding speeds, liquid instabilities can become important and may act to destabilize the weld through the production of a number of discontinuities that compromise weld quality (Albright and Chiang 1988a,b).

In practice, in industrial welding applications, productivity is improved by high weld speed. As an illustration, laser welding of tailor blanks for automotive applications is typically carried out at speeds of 6–15 m/min to maintain profitability. Although such high speeds are not possible in heavier-gauge welding, the trend always is to maximize weld speed while maintaining weld quality. The use of the data in Table 3.3 should be used only to define a possible range of operation for a given laser welding application.

An important consideration in any laser welding application is the sensitivity of the process to changes in system parameters. This is given by the slope of the response curve at the normal operating point. In general, this involves the derivative of the response function with respect to each of the system variables. From equation 1, $\Delta[P/vd]/\Delta v \cdot \alpha \cdot v^{-2}$; therefore, this quantity, which relates penetration depth and welding speed to laser power, is most sensitive to velocity fluctuations at slow speeds.

Since

$$\frac{\Delta d}{\Delta P} = \frac{1}{(av + b)},$$   (3.2)

where $a$ and $b$ are the parameters in the recursion fits, changes in penetration depth, $\Delta d$, are least sensitive to changes in laser power, $\Delta P$, at high welding speed.

With the large database available, it is not surprising that correlations have been identified between laser welding parameters and material properties. Some of these correlations are plotted in Figures 3.10 and 3.11. Metzbower (1993b) reported the following analytical fit to the data shown in Figure 3.10 for $CO_2$ laser welding of ASTM A36 steel:

$$d = \frac{0.10618P}{(KT_m)}\left[\frac{vb}{\kappa}\right]^{-1.2056},$$   (3.3)

where $d$ is given in meters, $v$ in meters per second, and $P$ in watts. The thermal conductivity K (W/m°k), thermal diffusivity $\kappa$ (m²/sec), and beam diameter $b$ (m) are included in this fit.

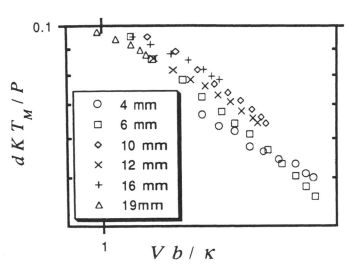

**Figure 3.10.** Dimensionless log-log plot of ASTM A36 laser beam welds. From Metzbower (1993).

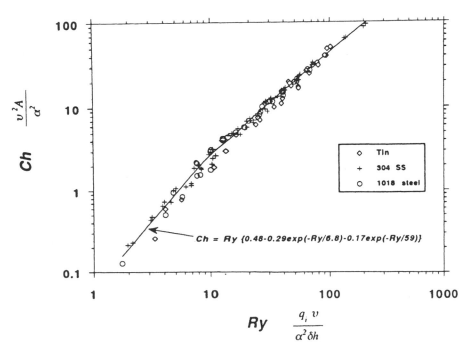

**Figure 3.11.** Dimensionless parameter model for partial-penetration continuous-power laser welding. From Fuerschbach (1994).

Fuerschbach (1994) reported an excellent correlation between the dimensionless parameters

$$Ry = \frac{P_i v}{\kappa^2 \Delta H_m} \tag{3.4}$$

and

$$Ch = \frac{v^2 A}{\kappa^2}, \tag{3.5}$$

where $P_i$ is the power into the workpiece, $\Delta H_m$ is the enthalpy of melting, and $A$ is the weld cross-sectional area, with $\kappa$ to be valued at the liquidus temperature. Laser welding data for autogenous welds in tin, 304 stainless steel, and 1018 steel are plotted in Figure 3.11 and are well correlated with these parameters.

The correlations described above, together with data in plots such as those in Figures 3.10 and 3.11, can be used in the initial selection of parameters for laser welding of a particular material. This can usually be reduced to a specification of the $P$, $v$, and $d$, that should result in an acceptable weld. Optimization of welding conditions, however, requires additional experimentation using the initial set of $P$, $v$, and $d$ coordinates as a starting point.

One approach to optimization is to construct a matrix of experiments to investigate the effect of relatively small changes near this initial condition. Taguchi methods may be used to optimize the experimental design (Watt and Uddin 1994; Yip, Man, and Ip 1996), but frequently other simpler methods may be acceptable (Dawes 1992). One technique is to classify welds as acceptable or nonacceptable within a matrix of experiments in which $P$, $v$, and other process variables are systematically varied by up to $\pm 20\%$ of the initial values. Penetration depth and weld shape may be used as criteria in the evaluation of these experiments. With pulsed lasers, similar experimental matrices can be developed, but a determination of optimized conditions may become difficult when such process variables as temporal pulse shape and duty cycle are included (McCay et al. 1990).

A summary of representative experiments that have been carried out in connection with the characterization and optimization of laser welding of metals is given in Table 3.4.

## 3.4   JOINT CONFIGURATIONS

Flexibility is one of the major attributes of laser welding systems and leads to a wealth of opportunities in joining materials in a variety of geometries. In most instances, laser welding can be carried out without the use of filler material, but in all cases the difference between a good weld and one that is unacceptable is the maintenance of fit up within the weld seam. This requirement results in tight toler-

**TABLE 3.4. Representative Experimental Studies of Laser Welding of Metals**

| Material | Laser | Power Level (kW) | Measured Parameter | Reference |
|---|---|---|---|---|
| 304 SS | $CO_2$ | 20 | Basic study | Locke, Hoag, and Hella (1972) |
| Rimmed steel | $CO_2$ | 5 | Basic study, joint configuration | Baardsen Schmatz, and Bisaro (1972) |
| 302 SS, 321 SS, Inconel, Monel, Ti | $CO_2$ | 0.25 | Basic study, low-power welding | Webster (1970) |
| 316 SS, 310 SS, Ducol W30 | $CO_2$ | 5 | Weld profile, tensible test | Willgoss, Megaw, and Clark (1979) |
| 304 SS | $CO_2$ | To 14.3 | Weld profile, penetration depth | Carlson and Gregson (1988) |
| Low carbon, 304 SS | $CO_2$ | 10, 20 | Effect of focal length, porosity | Funk et al. (1989) |
| 0.2% Carbon steel | $CO_2$ | 10 | Effect of f/number | Kaye et al. (1982) |
| HY-130 steel | $CO_2$ | 11 | Tensile strength, porosity | Stoop and Metzbower (1978) |
| HY-130, ASTM A36, A710 | $CO_2$ | To 12.25 | Penetration depth | Metzbower (1992) |
| 304 SS, mild steel | $CO_2$ | 5 kW pulsed | Effect of pulsing on penetration depth and bead width | Kimara, Sugiyama, and Mizutame (1986) |
| Austentitic stainless steel | $CO_2$ | To 9 | Microstructure | Kujanpaa and David (1986) |
| Zn-coated steel | $CO_2$ | To 8 | Joint design, melting, joining efficiency, corrosion rate | Akhter, Steen, and Cruciani (1988); Akhter et al. (1989a) |
| Coated steels | $CO_2$ | 2 | Effect of coating weight, finish, passivation | Waddell et al. (1987) |
| 304 SS | $CO_2$ | 0.2 | Thin sheet, high speed, stress and fatigue strength | Arai et al. (1987) |
| Coated, electron-plated steel | $CO_2$ | 5, Pulsed | Effect of frequency, duty cycle, volume of Zn evaporated per pulse | Heyden, Nilsson, and Magnusson (1989) |
| Stainless steel | $CO_2$, Nd: YAG | 6,2 Pulsed, CW | Comparison of laser sources, lap welds | Leong et al. (1993) |
| A1SI 316 | $CO_2$, Nd: YAG | 0.4, Pulsed | Comparison weld profile, cracking | Bagger, Nielsen, and Olsen (1993) |
| 1010 steel | Nd:YAG | 2.4 | Penetration depth | Scheuerman (1993) |
| Steel | Nd:YAG | 2 | Effect of shielding gas, porosity | Takano et al. (1990) |

*(continued)*

**TABLE 3.4.** (continued)

| Material | Laser | Power Level (kW) | Measured Parameter | Reference |
|---|---|---|---|---|
| 304 SS | Nd:YAG | 0.20 V, 0.4 pulsed CW + pulse wave | Thin sheet, bending test, effect of pulsing | Minamida et al. (1991) |
| BS 4360-43A steel | $CO_2$ | 1.2 | Underwater welding | Shannon et al. (1994a); Shannon, Watson, and Dean (1994b) |
| Al 5456 | $CO_2$ | 8 | Mechanical properties, elemental analysis | Moon and Metzbower (1983) |
| Aluminum alloys | $CO_2$, Nd: YAG | Various | Review | Thorstensen (1989) |
| Aluminum alloys, 5000, 6000 series | $CO_2$ | 3 | Weld profile, hardness, penetration depth, effect of shield gas | Behler et al. (1988) |
| Various Al alloys | $CO_2$ | 3.2 | Porosity, weld profile, hardness | Katayama et al. (1989) |
| Al 8090 | $CO_2$ | 2 | Weld width, penetration depth, focal position, lap welds | Kamalu, McDarmaid, and Steen (1991) |
| Al 2090 | $CO_2$ | 1.3 | Penetration depth, effect of surface preparation | Molian and Srivatsan (1988) |
| Various alloys | $CO_2$ | 5 | Effect of assist gas, focal position on weld profile | Matsumura et al. (1992) |
| Al 5000, 6000 series | $CO_2$ | 5–10 | Tensile strength, elongation, porosity, filler | Jones et al. (1992) |
| Al 2000, 6000, 7000 series | $CO_2$ | To 2.5 | Effect of filler, gas flow, lap welds | Bonello and Bailo (1993) |
| Al 5000, 6000 series | $CO_2$ | To 5.0 | Porosity, tensile strength | Rapp et al. (1993) |
| AC 120, Al 6009 | $CO_2$, CW, pulsed | 6 | Weld profile, effect of pulsing, tensile strength | Richter et al. (1993) |
| Al 2219 to Al 6061 | $CO_2$ | To 3.7, pulsed 5 kHz | Microstructure, hardness | Gopinathan et al. (1993a) |
| Al 5000, 6000 series | Nd:YAG | 0.4, 0.6 Pulsed | Microprobe studies, penetration depth, cracking | Cieslak and Fuerschbach (1988) |

*(continued)*

**TABLE 3.4.** (*continued*)

| Material | Laser | Power Level (kW) | Measured Parameter | Reference |
|---|---|---|---|---|
| Al 1050 | Nd:YAG | | Effect of pulse length, focussing on bead shape and penetration, lap welds | Watanabe and Yoshida (1990) |
| Al 5052, 5152 | Nd:YAG pulsed and CW | 0.7 | Penetration depth, bead width, cracking | Aruga et al. (1992) |
| Al 5052, 5083 | Nd:YAG | 4.2 | Penetration depth, porosity | Nishimi et al. (1996) |
| Al 5052, 5083 | Nd:YAG | 0.9–1.6 | Effect of focus, surface preparation | Xijing, Katayama, and Matsunawa (1997) |
| Al 5754-0, 6111-T4 | $CO_2$ | 3 | Elongation, microstructure | Venkat et al. (1997) |
| Various Al alloys | CO | 1.5 | Penetration depth, compare with $CO_2$ | Mehmetli, Takahoshi, and Sato (1996) |
| Magnesium alloy | $CO_2$ | 1.0 | Penetration depth, weld profile, microstructure | Maisenhalder, Chen, and Roth (1993); Chen et al. (1993) |
| Ti-6 Al-4V | $CO_2$ | 2–5 | Penetration depth, melting efficiency | Mazumder and Steen (1980) |
| Ti-6 Al-4V | Nd:YAG | 1.05 | Weld profile, penetration depth | Hoult (1990) |
| Copper, Ni-plated | Nd:YAG | 0.18 | Weld profile, strength | Hashimoto, Sato, and Niwa (1991) |
| Copper to steel | $CO_2$ | 2 | Penetration depth, weld profile, microstructure | Dell, Erba et al. (1986) |
| Copper to steel | $CO_2$ | 3.7 | Effect of beam position, cover gas, weld profile | Gopinathan et al. (1993b) |
| Aluminum to Cu | Nd:YAG | 0.15 | Tensile strength, use of filler, hardness, weld profile | Sarady, Lundquist, and Magnusson (1993) |
| Various dissimilar | | | Review | Sun and Ion (1995) |

ance on the geometry of the seam and the relation of the laser focus with respect to this seam.

Laser welding can occur only when the heating effect of the laser radiation extends well into the joint to be welded, and this leads to certain optimized approaches to interfacing between the laser beam and the parts to be welded together. A number of standard joint geometries are summarized in the following sections and in Figure 3.12. Unless the parts to be joined by laser welding are very thin, penetration rather than conduction welding conditions are preferred, although larger section materials also can be usefully welded via conduction under certain conditions. Spot welding as either individual or overlapping spots also can be substituted for seam welding in many of these joint configurations; in this case, weld penetration is determined by pulse energy, pulse length, and the number of overlapping pulses.

### 3.4.1  Butt Weld

This simple joint geometry is encountered in blanking operations, particularly with laser-welded tailor blanks (see Section 3.7). Primary fit up considerations include the need to minimize any gap between the sheets that is $>{\sim}0.05$ times the smallest thickness. For similar gauge welding, the laser beam should overlap both sides of the joint and should not wander off the centerline by more than $\pm 10\%$ of the beam diameter. For a typical beam diameter of 300 $\mu$m, this represents $\pm 30$ $\mu$m or ~1 thou. When welding dissimilar gauges or materials of different composition, the optimum focal position often is off the centerline of the joint. The butt configuration is preferred where possible when welding galvanized steel sheet because it allows the escape of zinc vapor from the joint.

The small amount of material available in the fusion zone during welding makes butt welding of thin sheets susceptible to pinhole formation. This is aggravated by the presence of any gap between the sheets. Full penetration must be maintained to obtain good mechanical properties from butt-welded seams. Clamping to maintain fit up is of utmost importance in butt welding. The sheets also have a tendency to missalign during welding due to thermal stresses and distortion; for this reason, long-seam butt welds in thin sheet materials often are tack welded before the seam weld is begun.

A backing plate with the same composition as the alloys to be welded can be used in butt welding of low-viscosity materials such as some aluminum alloys. The root of the weld extends into the backing plate to form a T joint. This prevents dropthrough at the root end of the weld and undercutting at the entrance to the weld.

### 3.4.2  Lap Weld

Lap welds are encountered in many applications involving the joining of thin sheets. Deep-penetration welds also are possible through thicker sections. Unless the position of the weld seam is critical, lap welds are forgiving of beam missalignment. Clamping is required, however, to obtain good bonding at the interface between sheets, although this clamping must include provision for a fine gap when lap welding

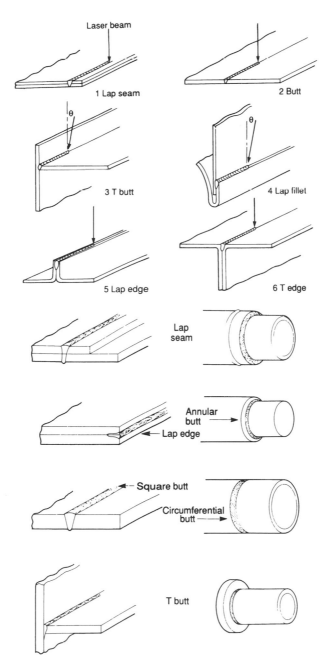

**Figure 3.12.** *Basic joint configurations for laser welding of sheets (top) and tubing (bottom). From Dawes (1992).*

galvanized steels or other coated materials with CW laser radiation. This can be accomplished by dimpling the sheet or adjusting the clamping jig to maintain a ~0.1-mm gap between sheets (Steen 1991, Dawes 1992). Laser knurling also has be shown to be effective (Forrest et al. 1997).

Pulsed laser lap welding of zinc-coated steel is possible without a gap if the pulse repetition frequency and pulse energy are optimized (Heyden et al. 1990). Figure 3.13 shows lap welds between galvanized steel sheets using pulsed $CO_2$ laser radiation. Some undercutting is present when no gap is used, but the overall weld quality is satisfactory. Heyden, Nilsson, and Magnusson (1989) reported that a volume of ~$10^{-3}$ mm$^3$ of Zn is removed per pulse, but this vapor can escape before the next pulse arrives, preventing the buildup of the pressure that causes gap separation. A complete discussion of the role of the gap in laser welding of Zn-coated steel is given by Akhter et al. (1988), Akhter and Steen (1990), and Bagger et al. (1992). These authors derive an expression for the ratio gap, *g,* to sheet thickness, *t,* where

$$\frac{g}{t} = \frac{A v t_{Zn}}{t^{3/2}}, \tag{3.6}$$

where v is the welding speed, $t_{Zn}$ is the thickness of Zn on each sheet at the interface, and *A* is an empirical constant. For $CO_2$ laser welding, $A = 18.25$ sec·m$^{-1/2}$. Values for *g/t* of ~0.2–0.3 are found to be consistent with good-quality welds.

A variety of techniques have been used to obtain the requisite gap during laser lap welds with shims (Rito et al. 1988, Akhter, Steen, and Cruciani 1988, Heyden, Nilsson, and Magnusson 1988) or prestamped projections (Petrick 1990, Piane et al. 1987, Spies and Thomas 1992) or via rolling during welding (Hanicke and Stradberg 1993). The Zn coating also may be removed before welding (Williams et al. 1993).

Under deep-penetration keyhole welding conditions, multiple sheets can be suc-

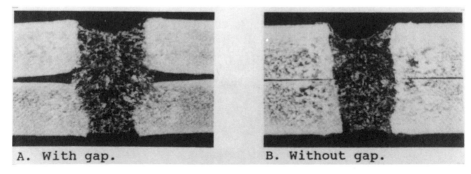

A. With gap.          B. Without gap.

*Figure 3.13.* Pulsed $CO_2$ laser welding of coated steel with (a) and without (b) gap. From Heyden, Nilsson, and Magnusson (1989).

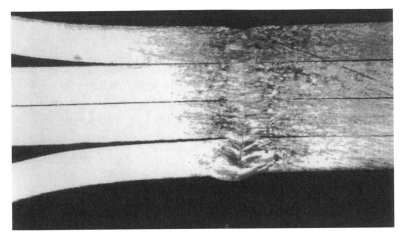

**Figure 3.14.** *Transverse cross section through a four-layer lap seam weld between steel sheets, each 0.7 mm thick, illustrating another example and advantage of the deep-penetration capability of laser welding. From Duhamel (1996).*

cessfully welded if appropriate clamping is used. Figure 3.14 shows such a weld extending through four 0.7-mm-thick steel sheets. When welding two or more sheets of different thicknesses, the preferred geometry has the thin sheet on top of the thick sheet. Additional configurations are described by Mombo-Caristan (1996a,b 1997).

### 3.4.3  Edge Welds

Two examples of edge welds are given in Figure 3.12, but other possibilities exist for this type of joint configuration. Through the careful choice of alignment, weld speed, and laser beam power, thin-gauge material can be penetration welded to a depth that is many times the sheet thickness. Misalignment in this configuration can result in burn through or lack of full penetration, and the overall strength of the join depends on the penetration depth.

Although this geometry can be more tolerant of gaps, good fit up is still necessary, and the maximum gap separation should not exceed 0.05 times the thickness of the thinnest sheet. Welding of galvanized sheet is facilitated by the presence of the edge, which allows zinc vapor to escape; however, deep-penetration welds may still require a gap to prevent the buildup of excess pressure. An example of an edge weld in automotive steel is shown in Figure 3.15.

### 3.4.4  T Butt and Lap Fillet Welds

Requirements for optimization of T butt and lap fillet welds was discussed by Dawes (1992). The laser beam must be incident at an angle of 7–10° unless the joint forms a corner when normal incident is possible. Focal position relative to the joint is critical if full penetration is to be obtained because beam reflection can

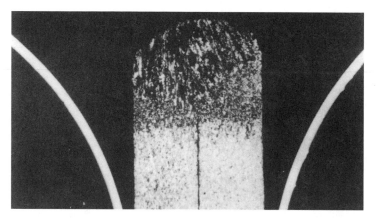

**Figure 3.15.** Flange joint welding of automotive body sheets (laser power, 4 kW; welding speed, 7 m/min). From Imhoff et al. (1988).

occur, which directs radiation away from the joint. Good weld strength is obtained only when both components are fully fused along the seam line. The maximum gap that can be accommodated typically is 0.05 times the thickness of the web plate, and the laser beam should align with the seam to within ~0.5 times the laser spot diameter. With heavy-gauge materials, T butt welds can be carried out on both sides of the joint. The fusion zone often is observed to curve away from the direction of the incident laser beam toward the joint that enhances the filling of the joint with the fusion zone. This reangling of the keyhole arises from thermal conduction and produces the effect known as ''beam skidding,'' or a ''skid weld.''

### 3.4.5   Flare Weld

This configuration (Figure 3.12) uses the high reflectivity of metals at oblique incidence to channel laser radiation to the root of a wedge formed between two sheets. With the correct polarization of the laser beam, incident laser power can be efficiently deposited at the point at which the two sheets come together (Behler et al. 1988a,c). Applications include the welding of pipes (Sepold, Rothe, and Teske 1987, Minamida et al. 1986, 1991) and the manufacture of plated metal sheets (Schulz and Behler 1989). Laser heating can be combined with electric resistance welding in this application for improvements in process efficiency (Minamida et al. 1991).

Flare welds are also useful in other joint configurations (Figure 3.16) at which joint geometry allows the laser beam to be ''trapped'' at the seam after reflection from the surrounding joint components. Because this method efficiently uses laser power, welding speeds can be high. One application involves the joining of coils by means of an elliptical lap weld (Figure 3.17).

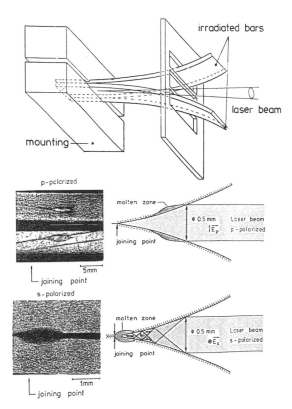

**Figure 3.16.** *Experimental set-up and evidence that s-polarization leads to deep penetration and a weld at the joining point is produced. Due to high absorption, a p-polarized beam does not weld the metal sheets; therefore, the irradiated bars are photographed side by side. From Schulz and Behler (1989).*

### 3.4.6  Narrow Gap Welding

An alternative to the use of laser radiation to form a keyhole to facilitate penetration welding is to create a physical gap between the parts to be welded that then acts to trap incident laser radiation and distribute it along the joint (Shewell 1977). With the correct gap separation, angle of incidence of the laser beam, and shaping of the sides of the gap, multiple reflections result in enhanced heating deep within the joint. This effect was investigated by Carlson and Gregson (1986), Lampa et al. (1995), and Milewski, Keel, and Sklar (1995). With gaps that have parallel sides, welding is improved at small values of gap separation ($\lesssim 0.1$ mm), and the weld has a flatter profile. Above a critical value, however, the welding process collapses because most laser radiation passes through the joint. This effect is less important for large sheet thicknesses or when the sides of the joint can be angled relative to the beam (Figure 3.18). Tapering of the sides of the joint over part or all of the sheet thickness also has been shown to be effective (Milewski et al. 1995).

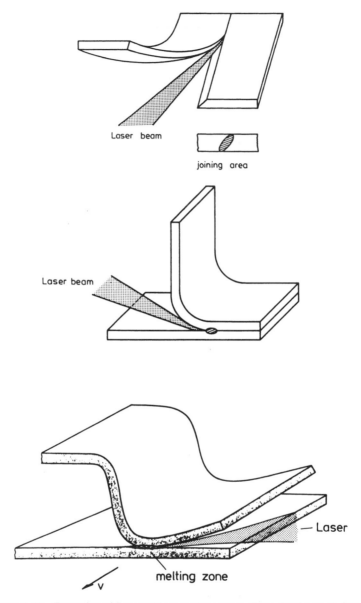

**Figure 3.17.** *Examples of flare welds in various joints. From Behler et al. (1988c).*

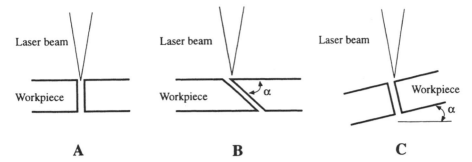

**Figure 3.18.** *Changing the beam/workpiece geometry from A to B or C will reduce gap sensitivity, although processing speeds may have to be reduced. From Lampa et al. (1995).*

### 3.4.7 Other Joint Configurations

Laser beam welding offers the design engineer a new technology for the joining of complex structures. Traditional joint designs that work well with arc or resistance welding techniques may not be optimized for laser welding. This question has been addressed (Larsson 1993, Ishihara, Troy, and Shirai 1993, Nilsson et al. 1993, Dawes 1992, Vastra and Diotalevi 1994, Walther 1994) and has led to a variety of innovative designs.

A primary limitation of conventional resistance welding, namely, the need to insert an electrode behind the part to be welded, is eliminated in laser welding and has created many new opportunities for welded joints in automotive manufacture. One of the earliest of these to be reported was the use of $CO_2$ laser radiation to form the ditch weld between the roof and side ring (Uddin et al. 1986). This is a standard stage in the assembly of vehicles (Larsson 1993). Ishihara et al. (1993)

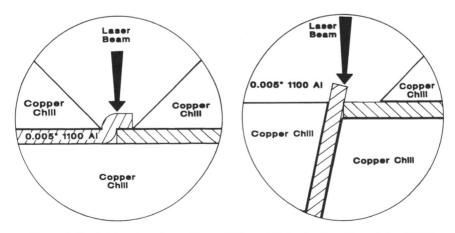

**Figure 3.19.** *Joint designs for welding of thin metal sheets. From Lingenfelter (1987).*

discuss how laser welding can simplify the assembly of door pillars and show how a new laser-welded pillar design cannot be assembled with conventional resistance joining techniques because it lacks access for the second welding electrode.

Joint design for laser welding of tubes is discussed by Nilsson et al. (1993) and Hayashi et al. (1996), whereas that for the welding of honeycomb structures is reported by Marsico and Kossowksy (1988), Oikawa et al. (1993), and Marsico, Denney, and Furio (1993). General considerations relating to the fit up of parts for laser welding are reported by Olsen (1994).

Problems associated with welding of thin–cross section materials (<0.1 mm) were addressed by Lingenfelter (1987), who report that standing edge and overlap joints are the best choices. Butt welds are less desirable. An example of two success-ful joint designs with thin Al 1100 material is shown in Figure 3.19.

An example of clamping and joint design for Nd:YAG laser welding of thin Ni-clad steel sheet is described by Newsome, Meyer, and Albright (1987). In this case, the weld was carried out from the steel rather than the Ni-clad surface.

## 3.5   SPOT WELDING

Spot welding was the first welding operation to be carried out with lasers. This initially was an application that involved ruby lasers, but the higher-pulse repetition rates and pulse-tailoring capabilities attainable with Nd:YAG and $CO_2$ lasers have meant that spot welding is a standard application for these devices. A summary of early work can be found in the articles by Cohen and Epperson (1968), Anderson and Jackson (1965), Gagliano and Zaleckas (1972), and Cohen, Mainwaring, and Melone (1969). A comprehensive treatment of Nd:YAG spot welding was reported by Nonhoff (1988).

The penetration depth, melt area, and porosity of spot welds can be a complicated function of laser pulse energy, intensity, and pulse shape (Watanabe and Yoshida 1991, Matsunawa et al. 1992, Bransch et al. 1991, 1992, Simidzu et al. 1992, Kim, Watanabe, and Yoshida, 1993, Kugler and Bransch 1993). An example from Wata-nabe and Yoshida (1991) of the dependence of spot weld cross section in lap welding of 304 stainless steel sheets is shown in Figure 3.20. The weld strength depends on the area of overlap between the sheets and was found to be greatest for laser intensities in the range of 0.2–0.4 $MW/cm^2$. For constant pulse energy, this defines an optimum pulse length. The effect of increasing the thickness of the top sheet in the lap weld also is shown in Figure 3.20. Under standardized conditions, increasing the thickness of the top plate was found to result in lower weld strength.

With constant pulse energy, the penetration depth often increases with a decrease in pulse duration. This corresponds to higher peak intensity and increased coupling but is accompanied by an increase in porosity and a reduction in weld quality. The result is that optimum pulse energy and incident laser intensity usually can be defined that combine weld penetration and weld strength. In practice, this also restricts the focal area to a well-defined range, as well as the location of the focus in relation to the surface of the top sheet. An example of the effect of pulse duration and energy

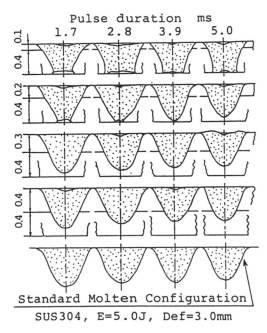

**Figure 3.20.** Nd:YAG lap welding of 304 stainless steel. Sheet thickness is given in millimeters. From Watanabe and Yoshida (1991).

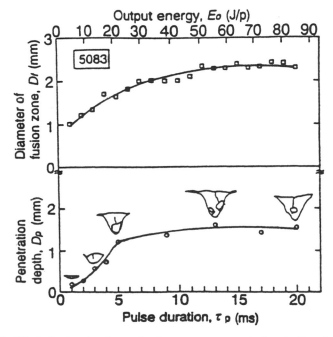

**Figure 3.21.** Effect of pulse duration and pulse energy on penetration depth and weld diameter for Nd:YAG laser (welds in Al 5083). From Matsunawa et al. (1992).

on penetration depth and weld morphology in Nd:YAG spot welds in Al 5083 is shown in Figure 3.21. Centerline cracking and porosity are evident in these welds, but they can be minimized through pulse shaping (Matsunawa et al. 1992, Matsunawa 1994). Tailoring of pulse temporal profile to minimize weld defects is discussed further in Section 4.8. A weldability test for Nd:YAG spot welding was described by Weeter, Albright, and Jones (1986).

## 3.6 MICROWELDS

Laser welding of fine wire and other small components has been an important application of lasers since their inception. These applications involve the use of pulsed laser radiation, with the earliest processing being carried out with ruby lasers (Gagliano and Zaleckas 1972). Joint design is important as is the placement of laser radiation on the components to be welded. In general, laser radiation must be applied to both components for optimized weld properties, although the exact location as

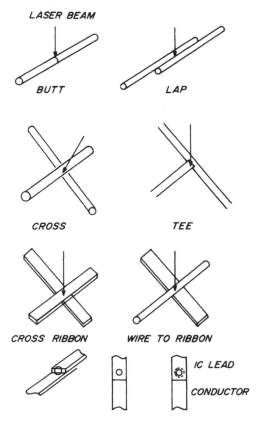

**Figure 3.22.** *Laser microweld configurations.*

well as optimal pulse energy is different for each application. A recent discussion of the techniques used to achieve good weld properties has been reported by Wojcicki and Pryputniewicz (1996). Reviews of microwelding joint design and weld properties under various conditions were reported by Gagliano, Lumley, and Watkins (1969), Gagliano and Zaleckas (1972), Duley (1976), Rykalin, Uglov, and Kokora (1978), Shewell (1977), and Mazumder (1983). Some common joint configurations are shown in Figure 3.22.

## 3.7 MECHANICAL PROPERTIES

The creation of a weld between two components inevitably results in the modification of mechanical properties such as hardness, toughness, tensile strength, fatigue strength, and formability. The metallurgical basis for these changes is discussed in Chapter 5 with an emphasis on the variation in hardness across fusion- and heat-affected zones in laser welds. The present section outlines some of the data available on the static and dynamic strength of laser welds in metals. Mechanical properties of laser welds in tailor blanks are discussed in a later section.

### 3.7.1 Static Strength

Transverse and longitudinal tensile tests are commonly used to evaluate the strength and elongation of laser welds in metals. Most studies have centered on the automotive steels used in tailor blanks (Kitani, Yasuda, and Kataoka 1995, Baysore et al. 1995, Yoshida et al. 1995, Fukui et al. 1996, Saunders and Wagoner 1996). Data also are available on some Al alloys (Rapp et al. 1993; Starzer et al. 1993; Duley 1994; Venkat et al. 1997).

In steels, tensile failure usually occurs in the base metal in transverse tests due to increased hardness in the weld and heat-affected zones. Failure in the weld, however, is present in longitudinal tests with cracks appearing transverse to the weld bead.

With Al alloys, the failure mode depends on alloy type and temper, as well as on welding conditions and whether filler wire is used. Under transverse tensile testing, $CO_2$ laser-welded Al 7075-0 fails in the base metal, whereas 6060 in the T4 or T6 temper fails within the fusion zone. The addition of filler has been shown to shift crack failure from the fusion zone to the heat-affected zone (Starzer et al. 1993).

In low-carbon automotive steels, laser-welded joints are found to have tensile strengths that are comparable to that of the base material and elongations up to ~90% of base material. Fukui et al. (1996) reported that tensile strength and elongation of laser-welded low-carbon steel blanks are greater than those of mash/seam-welded samples.

Little data are available for laser-welded joints in Al alloys. Under optimized welding conditions, however, in certain alloys, tensile strengths and elongations

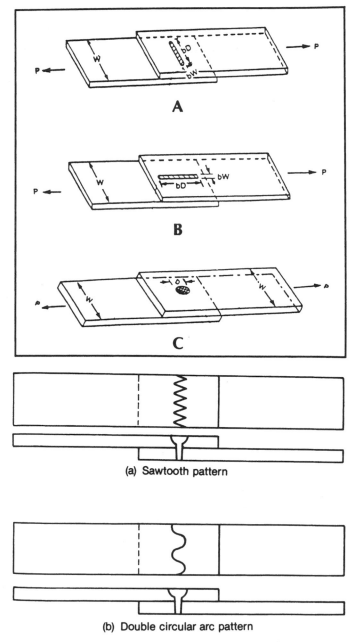

**Figure 3.23.** *Joint designs for lap welds. (Top) From Wang and Ewing (1991). (Bottom) From Albright, Hsu, and Lund (1990).*

comparable to those of the base metal can be obtained (Duley 1994; Venkat et al. 1997).

### 3.7.2 Fatigue Strength

Although there have been numerous studies of the static strength of laser-welded joints, the application of laser welding techniques in the assembly of automotive components requires testing under dynamic loading conditions. Several experimental studies have addressed the question of the fatigue strength of laser spot and stitch welds in comparison to resistance spot welds (Albright, Hsu, and Lund 1990, Wang and Ewing 1991, 1994a, 1994b, Wang and Davidson 1992, Fraser and Metzbower 1983, Shimbo, Ono, and Kabasawa 1993, Flavenot et al. 1993, Guglielmino et al. 1993, Wang 1993), with generally positive results.

A variety of lap weld designs have been evaluated that extend from simple transverse and longitudinal welds to circular and sawtooth patterns (Figure 3.23). Stitch as well as continuous seam welds also have been studied with various joint clearances

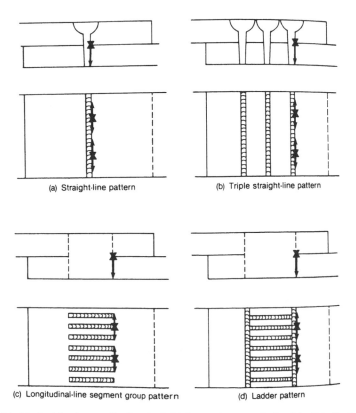

(a) Straight-line pattern  (b) Triple straight-line pattern

(c) Longitudinal-line segment group pattern  (d) Ladder pattern

**Figure 3.24.** *Typical fracture paths for straight-line and segmented welds. X indicates fracture initiation site. Arrow indicates direction of propagation. From Albright, Hsu, and Lund (1990).*

in coated as well as in noncoated steels. When compared on the basis of weld volume to resistance spot welds, simple transverse laser welds were found to have ~40% higher fatigue strength at $2 \times 10^6$ cycles (Wang and Ewing 1991) in SAE 1008 Al-killed sheet steel. In another study, the fatigue resistance of continuous transverse laser welds was found to be sensitive to the presence of a gap between the sheets, with a 10% gap decreasing the fatigue strength at $10^6$ cycles by ~29% (Wang and Ewing 1994b). Galvanization had little affect on fatigue strength under optimized welding conditions.

Other variables that were found to decrease fatigue life were increasing bead width, underfill, incomplete bead penetration, and increasing sheet thickness. Fatigue failure has been found to be initiated from microcracks at the edge of the heat-affected zone, particularly those at the beginning and end of the weld. These microcracks are present before testing and propagate along the edge of the weld bead in the heat-affected zone. Some examples of crack propagation paths for various weld configurations are shown in Figure 3.24.

The placement and shape of laser weld tracks are important if optimized weld fatigue strength is to be obtained. A good analysis of this design problem in relation to laser welding of T joints for automotive applications is outlined by Wang and Ewing (1994a).

## 3.8   TAILOR BLANKS

Butt welding of sheet steel to form what is known as a "tailor blank" was invented in 1964 by Kerby (1964). Welding was initially carried out using an electron beam (Hinricks et al. 1974), but this has been largely replaced by laser beam welding. Laser-welded tailor blanks (LWTBs) are components of most automobiles, and the market for these components represents one of the primary applications of lasers in production operations (Figure 3.25).

The tailor blank concept is simple and involves the assembly of individual flat blanks with different gauges and compositions to create a sheet with the combined characteristics of the individual components. The resulting tailor blank is stamped to yield the final three-dimensional component. This part can be engineered to have heavy-gauge high-strength material only in the area required, while lighter-gauge material is maintained for weight reduction over other parts of the structure. Similarly, coated steels can be incorporated into that part of the component exposed to a corrosive environment, while uncoated material is retained for other, less-exposed locations.

There are many potential advantages of the use of tailored blanks for cost savings and weight reduction. The joining of blank components by laser welding, while meeting the specifications for weld integrity and subsequent stamping performance, is, however, a major technological challenge. Meeting this challenge has opened up a large market for laser welding technology. Some advantages of LWTBs and attendant technological advantages and disadvantages are summarized in Table 3.5 (see also DiPietro 1992 and Shibata 1996).

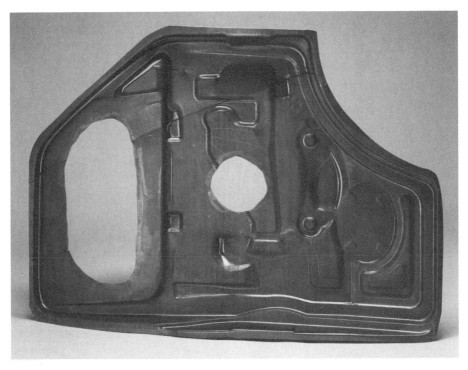

**Figure 3.25.** *Laser-welded tailor blank after stamping.*

**TABLE 3.5. Advantages of LWTBs and Requirements of Process Variables**

Advantages
Reduced number of components
Reduced weight
Lower assembly cost
Lower tooling and press shop costs
Optimized mechanical integrity and rigidity
Optimized corrorison resistance
Improved crash resistance and fatigue behavior
Process Requirements
Joint fit up critical
Edge preparation
Uniform penetration
Low porosity, no pinholes
Reproducibility

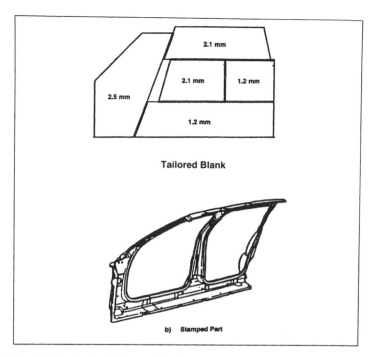

**Figure 3.26.** *Cadillac body side frame that is one of the most difficult to produce using laser beam blank welding. From Koons and Roessler (1994). (Reprinted with permission from IBEC '94. Copyright 1994 Society of Automotive Engineers Inc.)*

An example of the implementation of the LWTB concept in the manufacture of the Cadillac deVille-body in white is shown in Figure 3.26. The body side frame is initially assembled from five blanks with a range of thicknesses. The resulting stamped component optimizes strength and rigidity without the need for additional reinforcements (Irving 1991, Chalmers 1994). The annual volume of steel blanks is projected to grow into the $100–200 \times 10^6$ range by the early 2000s. An additional advantage of LWTBs involves the reduction in scrap produced through nesting of small components during initial blanking from the steel coil. Total scrap rates can be reduced to <10% with careful optimization of material use (Figure 3.27). Estimates predict up to 18 LWTBs per vehicle, which would represent ~20–25% by weight of the total steel in a typical automobile (Corrodi 1996).

**TABLE 3.6. Laser-Welded Blank Configurations and Applications in Automotive Body in White**

| Configuration | Application |
| --- | --- |
| Single straight seam | Door inner, center pillar, near deck lid |
| Multiple straight seams | Body side, frame rail, floor pan |
| Nonlinear seam | Shock tower, door inner |

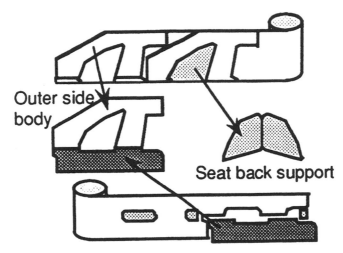

**Figure 3.27.** *Nesting of blank components in coil to produce low scrap rate. From Shibata (1996).*

Various joint configurations are possible depending on the part shape and function. A classification sequence was outlined by Corrodi (1996) and is listed in Table 3.6. Although many welds in LWTBs are linear (but not necessarily parallel), additional flexibility is obtained by using curved welds (van de Hoeven et al. 1996). Advantages include enhanced formability, elimination of reinforcements, and additional weight reduction. Disadvantages include edge preparation, clamping, and fit up to the desired precision. A typical tailor blank application in steel would involve the welding of 0.8-mm galvanized steel to 1.5-mm cold rolled high-strength steel. With a 6-kW $CO_2$ laser, the welding rate would be 6–8 m/min under the conditions summarized in Table 3.7.

The acceptance of LWTBs is subject to a wide range of validation criteria; these are summarized in GM 4485M as an engineering standard for weld specifications

**TABLE 3.7. LWTB System Parameters**

| | |
|---|---|
| Material | 0.8/1.5 mm galvanized/high-strength steel |
| Laser | 6 kW, $CO_2$ |
| Speed | 6–8 m/min |
| Focal length | 200–300 mm, parabolic mirror |
| Shield gas | Helium |
| Cover gas (root of weld) | Argon |
| Focal spot size | 0.2–0.5 mm |
| Fit up | <0.1 mm |
| Edge straightness | ≤±0.1 mm over 2 m |
| Shear-to-break ratio | 60/40 or better |

**TABLE 3.8. Criteria for Acceptance of Laser Welds as Specified in GM 4485M**

Concavity
Continuity
Convexity
Cupping tests (Olsen)
Cracks
Microhardness
Microstructure
Mismatch
Pinholes
Slag
Strength
Undercut

in laser welds. Categories of criteria are listed in Table 3.8. One of the most stringent of these criteria is that of mismatch that is obtained through measurement of the relative location of the two flush sides to the weld. The allowance in mismatch is 10% of the thickness of the smallest gauge material if <1 mm and 20% of this thickness if this exceeds 1 mm. An example of an acceptable weld between 0.8- and 1.5-mm materials is shown in Figure 3.28.

At the high welding speeds used in LWTB processing, the heat-affected zone is narrow, with typical widths in the 0.2–0.4-mm range. Microhardness increases in

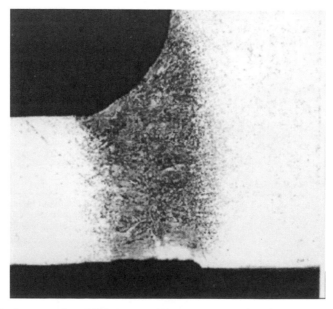

**Figure 3.28.** Cross section of $CO_2$ laser weld between 0.9- and 1.9-mm steel showing acceptable mismatch conditions. From Carter and Guastaferri (1996).

this region and maximizes in the center of the fusion zone, where it may be three to four times that of the base metal. This increase in hardness can influence formability but also means that failure under tensile test conditions occurs primarily in the base metal (Saunders and Wagoner 1996) rather than in the weld itself. Formability is measured using a dome test, and fracture heights are typically 85–90% of parent material. Draw depth is not severely affected by the presence of a laser weld, but care should be taken to locate the weld line away from regions of highest strain. Tensile test data and data on formability and fatigue strength in LWTBs are given by Natsumi et al. (1992), Baysore et al. (1995), Kitani et al. (1995), Yoshida et al. (1995), Fukai et al. (1996), and Saunders and Wagoner (1996). The fatigue limit of laser welds has been found to be comparable to or slightly higher than that of the base metal (Fukai et al. 1996). The effect of filler material on this and other mechanical properties was investigated by Kitani et al. (1995).

A major disadvantage and limitation of the LWTB technique is the need for the maintenance of high tolerances on sheet fit up and gap separation. This requires precision shearing of sheets before welding and a rigorous and precise damping system. Gaps as small as 0.1 mm can seriously influence the integrity of laser welds through lack of penetration and the formation of pores. Several techniques have been proposed to reduce or eliminate this problem (Table 3.9), but none of these alternative approaches are completely satisfactory for the reasons given in the table.

Although much LWTB processing has involved the use of $CO_2$ laser radiation, the Nd:YAG laser is an excellent alternative source. YAG laser welding offers the flexibility of fiber-optic delivery and increased coupling of laser radiation into the workpiece. Preliminary results with YAG laser welding of steel blanks with laser powers of 2–3 kW have demonstrated that welding of steel blanks can be carried out at speeds of >7 m/min. The relatively large spot diameter (typically 0.6 mm) results in a high tolerance for poor fit up conditions and minimizes the requirement for precision shearing. An attractive advantage of welding with Nd:YAG laser radiation compared with $CO_2$ laser welding is the elimination of the need for He as a cover gas to reduce plasma formation.

**TABLE 3.9. Methods of Reducing Edge Preparation Requirements for LWTBs**

| Technique | Disadvantage |
|---|---|
| Beam oscillation* | Complex optics |
| | Slower speed |
| Rolling of joint† | Slower process speed |
| | Increased mechanical complexity |
| | Reduced profile |
| Lap weld‡ | Increased mismatch |
| | Complex clamping |
| Laser/arc combination§ | Additional system complexity |

* From Kitani, Yasuda, and Kataoka (1995).
† From Wildmann, Urech, and Freuler (1996).
‡ From Mombo-Caristan (1996).
§ From Beyer et al. (1994a).

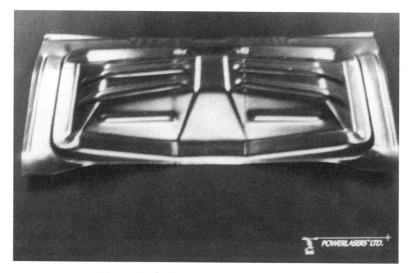

**Figure 3.29.** Stamped tailor blank in Al 5744.

Even though steel blanks dominate the market, there is much interest in laser welding of Al alloys for similar applications in the manufacture of automotive components. Duley (1994) and Adair (1994) reported on the successful welding of Al 5754 in the tailor blank configuration using $CO_2$ laser radiation. Autogeneous butt welds between various gauges of Al 5754 and between Al 5754 and Al 7075 have been obtained, and blanks formed from these materials have been successfully stamped (Figure 3.29). Table 3.10 lists some mechanical data for laser-welded Al 5754 material. Figure 3.30 shows etched cross sections of butt welds in Al 5754.

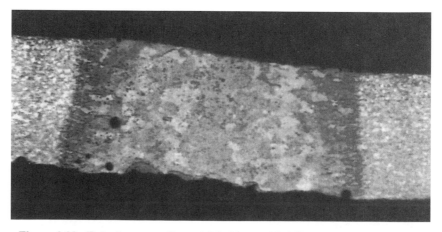

**Figure 3.30.** Etched cross sections of $CO_2$ laser-welded Al 5754. From Duley (1994).

**TABLE 3.10. Mechanical Test Data for Laser-Welded Al 5754**

| | Transverse Tensile Tests†‖ | | | | | |
|---|---|---|---|---|---|---|
| | Ultimate Tensile | | Yield Strength | | %<br>Elongation‡ | Formability§ |
| | Maximum | Minimum | Maximum | Minimum | Minimum | Minimum |
| Parent material* | 235 | 205 | 115 | 95 | 19 | 1.9 |
| Laser welded | 235 | 185 | 103 | 80 | 13.5 | 1.9 |
| | Longitudinal Subsize Tensile Tests‖ | | | | | |
| | Ultimate Tensile | | Yield Strength | | %<br>Elongation | Formability |
| | Maximum | Minimum | Maximum | Minimum | Minimum | Minimum |
| Laser welded | 270 | 245 | 128 | 110 | 18 | 2.0 |

* From Alcan International specification for cut-to-length sheet.
† All strengths are given in MPa.
‡ % Elongation is at fracture.
§ Formability is calculated at U.T.S/Y.S.
‖ Tests conform to ASTM Test Methods for Tension Testing of Metallic Materials (E 8M-93).
From Duley (1994). ASTM, American Society for Testing of Materials; U.T.S., Ultimate Tensile Strength; Y.S., Yield Strength.

These welds exhibited elongations of 10–16% in tensile tests. Similar results were reported by Venkat et al. (1997). Reviews of laser welding of Al alloys have been given by Rapp et al. (1995) and Dausinger et al. (1997).

The joining of steel and Al would provide additional flexibility in automotive manufacture, but the welding of these materials is difficult because the resulting fusion zone contains Al-Fe intermetallic compounds. This leads to a brittle weld and poor performance in forming tests. A novel approach that reduces this problem was described by Polzin et al. (1995) and Jaroni et al (1996). This technique involves a laser-activated cold rolling bond between steel and Al to form an overlap seam with a width of 3–10 mm. Formability tests show that this type of join is able to perform well in bending tests, but problems of cracking exist in deep drawing tests. Part of this difficulty may arise because of the cold rolling process, which leads to cold deforming before drawing. This type of lap join also has been subjected to corrosion testing (Jaroni et al. 1996), with acceptable results.

## 3.9 LASER WELDING WITH FILLER MATERIAL

With thick section welds or welding where gaps are present, filler wire or powder is used as an auxilliary source of material to fill the gap and facilitate welding with laser radiation. With Al alloys, some alloys have a tendency to lose volatile alloy components and may exhibit solidification cracking due to metallurgical changes in

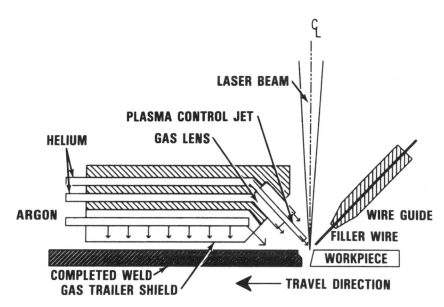

**Figure 3.31.** *Schematic of region of interaction of the gas management system, wire delivery system, and laser beam delivery system. From Carlson and Gregson (1986).*

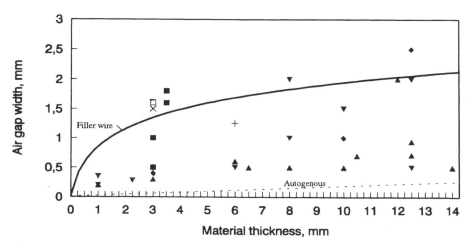

**Figure 3.32.** *Air gaps that can be filled with filler wire feeding as a function of material thickness (from experiments). ×, laser power of 1.9 kW; ●, laser power of 3.0 kW; +, laser power of 5.0 kW; ◆, 2 kW < laser power < 5 kW; ■, 5 kW < laser power < 10 kW, ▼, laser power of >10 kW. From Salminen, Kujanpaa, and Moisio (1994).*

the fusion zone. This situation often can be improved by the use of a filler wire of the appropriate composition.

The general configuration for laser welding with a wire feed system is shown schematically in Figure 3.31. The filler wire is introduced at the leading edge of the keyhole at an angle of ~45° and intersects the laser beam close to its focus at the middle of the joint. When the diameter of the filler wire exceeds the gap separation, the wire can slide along the top of the joint and interfere with the propagation of laser radiation to the root of the weld, resulting in poor penetration. When the gap width increases, the filler wire may not provide sufficient material to completely fill the weld bead. Both of these problems can be minimized through the use of a seam vision system with adaptive control over wire feed rate.

The ability of filler material to extend laser welding to larger gap separations and accommodate poor joint fit up is illustrated in Figure 3.32, obtained from the

**TABLE 3.11. Summary of Experimental Studies on Laser Welding Using Filler Materials**

| Metal | Filler | Gap (mm) | Thickness (mm) | Laser | Reference |
|---|---|---|---|---|---|
| Steel, low alloy | 1.2 mm, MIG wire | <2 | 8 | 5 kW $CO_2$ | Watson and Dawes (1985) |
| Steel, 0.2% C | 1.6–2 mm, 4GW-11 | 2–8 | 13.8 | 10 kW $CO_2$ | Arata et al. (1986) |
| Steel, ASTM A-36 | 1.1 mm, AWS 705-6 MIG wire | <0.94 | 12.7 | 8 kW $CO_2$ | Carlson and Gregson (1986) |
| Steel, BS1501-224-490B | 1.6 mm, flux cored | 0.75 | 13 | 9.2 kW $CO_2$ | Stares, Apps, and Megaw (1987) |
| Steel, St 37 | 0.8–1.0 mm MIG wire SG2 | <1.9 | 6 | 5 kW $CO_2$ | Salminen, Kujanpaa, and Moisio (1994) |
| Steel, low, ultralow carbon | 0.9 mm GMAW, ultralow carbon steel, high-Ni alloy | 0.2 | 0.8 | 5 kW $CO_2$ | Kitani, Yasuda, and Kataoka (1995) |
| Castiron, spheroidal graphite | 0.8 mm Ni | 0–0.5 | 6–9 | 8–22 kW $CO_2$ | Dilthey and Shu (1993) |
| Al 5083, Al 5754 | Strips, 0.07–1.0 mm Al-Mg, Al-Si | 0.07–1.0 | 11 | 156 W pulsed Nd:YAG | Junai et al. (1990) |
| Al 5083 | 1.0 mm Al 5056A | | 6 | 10 kW $CO_2$ | Jones et al. (1992) |
| Al 6060, Al 6082 | Si, AlSi-12 powder, AA 4047, 0.8–1.0-mm wire | | 4 | 2.5 kW $CO_2$ | Starzer et al. (1993) |

summary of Salminen, Kujanpaa, and Moisio (1994). The effect of filler is most dramatic in welding materials with thicknesses of ≲2 mm, at which gaps of up to ~50% of the sheet thickness can be accommodated. In heavier-gauge materials, welding with gaps of up to 2 mm is possible.

A quantitative investigation into the effect of filler wire on heat transfer during laser welding with a gap was reported by Carlson and Gregson (1986). Reflection of incident laser radiation by the filler wire and its effect on decreasing power deposition in the joint during welding can be significant and increase with wire feed rate and changes in laser spot size at the focus (Salminen et al. 1996).

Arata et al. (1986) carried out a detailed study of the effect of feed rate, gas assist flow, and welding speed on multipass $CO_2$ laser welding of low-carbon steel with the use of filler wire. They found that with filler wire, a 50-mm-thick plate can be butt welded with five passes at 10 kW. The welding speed was 0.3–0.5 m/min. Optimum weld properties were obtained at a welding speed of 0.4 m/min and a wire feed rate ~10 times this value.

The mechanical properties of low- and ultralow-carbon steel welded with $CO_2$ laser radiation and a wire filler were reported by Kitani, Yasuda, and Kataoka (1995). The addition of ultralow-carbon steel wire or high-Ni wire was shown to improve formability of low-carbon steel sheet. Welded joints containing weld metal of less than HV250 hardness showed improved formability in Erichsen tests.

With Al alloys, another role of the filler wire is to replace volatile alloy elements such as Mg and Si to minimize porosity and enhance weld strength. A study of the effect of the addition of Al 4047 filler wire on mechanical properties of laser-welded Al 6060-T4 was conducted by Starzer et al. (1993). The beneficial effect of filler material was found to increase with feed rate for rates of up to 2.5 m/min when welding at 3 m/min in 4-mm-thick sheets. Additional studies of the role of filler material in laser welding of Al alloys are summarized in Table 3.11.

# 4

## Heat Transfer and Modelling in Laser Welding

### 4.1 INTRODUCTION

The coupling of laser radiation into a metal to produce the localized heating required for spot or seam welding involves a delicate balance among many parameters. Some of these parameters, such as laser intensity, pulse shape, and beam polarization, are under the control of the operator, whereas others, such as metal reflectivity, thermal conductivity, and heat capacity, are not. Optimization of laser welding involves defining a set of experimental conditions that lead to stable and reproducible welding conditions and monitoring these conditions for quality assurance and possible real-time adaptive control.

In practice, laser welding of metals is an inherently unstable process due to such variables as fluctuations in laser output, plasma interference effects, and dynamic instabilities in liquid and vapor flow during welding. All of these parameters can influence the instantaneous power delivered to the workpiece, compromising weld properties and the mechanical integrity of the final join.

Understanding the nature of coupling between an incident laser beam and a metal under welding conditions provides insight into ways in which laser welding can be optimized. It also may suggest new directions for improvements in laser welding technology and how welding defects can be minimized. This insight can be obtained through an analysis of the sequence of processes that connect freely propagating laser radiation to the production of a weld pool and eventually a weldment.

The complexity of the interactions involved in this process and the subsequent behavior of liquids generated in a solid by incident laser radiation preclude a completely rigorous simulation of laser welding. As a result, a number of approximations are adopted in calculations of heat and mass transfer during laser welding to simplify this complex problem. In reality, the input of accurate experimental data on the physical and metallurgical properties of metals under the conditions encountered during laser welding is usually limited, and this immediately compromises the accuracy and relevance of theoretical predictions. However, despite these limitations, heat and mass transfer calculations are useful in quantifying the laser welding process and in suggesting how changes in process parameters may influence weld properties.

This section traces the sequence of steps involved in laser welding of metals using a simple model for beam absorption in conduction and penetration welding regimes. Heat transfer is discussed under both conditions and the results of simulations of melt flow in a laser-generated weld pool are summarized. In all cases involving penetration welding, the formation and properties of the laser keyhole are of prime importance, and these are discussed at length.

## 4.2  ABSORPTION OF LASER RADIATION

Laser radiation incident on the surface of a metal is absorbed by electrons. An electron that absorbs a laser photon makes a transition from one continuum state, $E_i$, to another state, $E_f$, with $E_f - E_i = h\nu$, the energy of the laser photon. With this excess energy, this electron is out of equilibrium and rapidly gives up this energy through collisions with other electrons and with lattice phonons. The short period of time, $\tau = 10^{-14}$–$10^{-15}$ seconds, over which this occurs after absorption of a laser photon ensures that the electron gas within a metal never becomes superheated under the irradiation conditions that prevail in laser welding (incident intensity, $<10^7$ W/cm$^2$). As a consequence, absorption of laser photons may be considered to instantaneously deposit energy at the site at which absorption occurs. In metals, this corresponds to a depth $\delta$ where

$$\delta = \frac{\lambda}{4\pi k}.$$ 

(4.1)

In equation 4.1, $\lambda$ is the laser wavelength, and $k$ is the imaginary part of the refractive index.

$$m = n - ik,$$ 

(4.2)

where $n$ is the real refractive index. At $\lambda = 1$–$10$ $\mu m$, $k$ for metals typically is $1$–$10$, so $\delta \sim 10^{-5}$–$10^{-6}$ cm. If $S$ is the area of the laser beam on the metal surface, then only electrons within the volume $V \sim S\delta$ absorb laser photons. These electrons will have speeds of $v \sim 10^8$ cm/sec and will lose their excess energy over a distance of $\ell \sim v\tau \sim 10^{-6}$ cm through collision with other electrons. Because $\ell \leq \delta$, the energy absorbed from the laser beam is deposited within a distance $\delta$ from the surface. The heat source on absorption of Nd:YAG or CO$_2$ laser radiation by a metal therefore can be considered to be localized at the surface.

For electromagnetic radiation at normal incidence on an opaque surface such as that of a metal, the absorptivity, $A$, is calculated as:

$$A = \frac{4n}{(n + 1)^2 + k^2}$$ 

(4.3)

**TABLE 4.1. Absorptivity (A) of Several Metals at 1.06 and 10.6 μm**

| | A | |
| --- | --- | --- |
| Metal | 1.06 μm (Nd:YAG) | 10.6 μm (CO$_2$) |
| Al | 0.06 | 0.02 |
| Cu | 0.05 | 0.015 |
| Fe | 0.1 | 0.03 |
| Ni | 0.15 | 0.05 |
| Ti | 0.26 | 0.08 |
| Zn | 0.16 | 0.03 |
| Carbon steel | 0.09 | 0.03 |
| Stainless steel | 0.31 | 0.09 |

Data are for room temperature.
From Duley (1983) and Xie et al. (1997).

independent of polarization (Born and Wolf 1975). Because both $n$ and $k$ are usually $\gg 1$ for metals at infrared wavelengths, $A$ is small (Table 4.1), and therefore only a small fraction of incident laser radiation is initially absorbed.

The absorptivity may be enhanced by a variety of factors, including temperature, surface roughness, oxidation, and changes in morphology (Duley 1976, 1983). It also changes at the melting point. The effect of an increase in absorptivity is to accelerate the deposition of energy from an incident laser beam, which in turn can result in a singularity in the overall heating process at specific combinations of laser intensity and irradiation time. This effect can be useful in tailoring laser pulse profiles for particular welding operations.

An example of this discontinuity in absorptivity at the melting temperature for Al and Cu is shown in Figure 4.1. These data show that $A$ for 10.6 = μm radiation increases by a factor of ~3 at the melting temperature. This follows a steady increase in $A$ from room temperature to the melting point. These results are for clean metal surfaces in a good vacuum; values of $A$ for surfaces in the presence of air will be significantly affected by oxidation (Duley 1983). The effect of an oxide layer is to introduce a thin dielectric film at the metal surface. This film can act as an interference filter with maxima and minima in $A$ depending on film thickness, $d$, and composition. In general, the film must have a optical thickness of $nd \geq \lambda/4$ for interference effects to be significant. Because $n \sim 2$–$3$ for common oxides, films with thicknesses as small as $d \sim 1$–$2$ μm will exhibit interference effects at 10.6 μm.

Thinner films also can enhance coupling of laser radiation to the surface if they are strongly absorbant. The absorption coefficient

$$\alpha = \frac{4\pi k}{\lambda} \tag{4.4}$$

can be as high as $2$–$5 \times 10^3$ cm$^{-1}$ at $\lambda = 10$ μm for strongly absorbing oxide films. If $I_0$ is the laser intensity at the surface of the film, then the intensity at the depth $z$ is

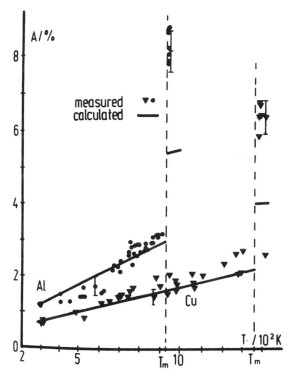

**Figure 4.1.** *Temperature dependence of the absorptivity, A, at 10.6 μm for Al and Cu surfaces in vacuum. Note the discontinuity at the melting temperature. From Bruckner, Schafer, and Uhlenbusch (1989). (Reprinted with permission from Journal of Applied Physics. Copyright 1989, American Institute of Physics.)*

$$I(z) = I_0 e^{-\alpha z} \tag{4.5}$$

$$\cong (1 - \alpha z)I_0; \quad \alpha z \ll 1. \tag{4.6}$$

With $\alpha = 5 \times 10^3$ cm$^{-1}$ and $z = 0.2$ μm $= 2 \times 10^{-5}$ cm, $\alpha z = 0.1$ and $I(z) = 0.9\, I_0$. The film then absorbs $0.1\, I_0$ W/cm$^2$ under these conditions. This heat will be rapidly conducted to the underlying metal surface, enhancing the overall efficiency of coupling of laser radiation into the metal.

Oxidation of a heated metal surface occurs rapidly at the temperatures reached during laser welding. This effect can be minimized through the use of an inert shielding gas such as He or Ar but likely occurs to some extent even with shielding. Because the enhancement in laser coupling provided by an absorbing thin film is important primarily in the initial heating phase, these oxide layers will not have much effect during welding but can influence the threshold intensity for surface melting and vaporization.

Surface roughness can have an important effect on laser coupling when the root-mean-square roughness $\sigma \geq \lambda$ (Oqilvy 1987). The reflectivity $R*$ of an opaque surface at normal incidence becomes

$$R* = R \exp -(4\pi\sigma/\lambda)^2, \tag{4.7}$$

where $R$ is the normal incidence reflectivity of the smooth surface. The reduction in reflectivity arises from a diffusely scattered component that redirects incident laser radiation at a range of angles away from the normal to the surface.

Although $A = 1 - R$ for an opaque surface, it is not necessarily true that $A* = 1 - R*$ for a rough surface. This occurs because the reduction in $R$ as the result of surface roughness is due to a redirection of radiation from the specular angle rather than to an increase in the absorptivity of the surface.

## 4.3  THRESHOLD FOR CONDUCTION WELDING

Under conduction limited conditions, the onset of surface melting can be estimated from the simple model shown in Figure 4.2. The temperature at the center of the beam focus ($r = 0$) is (Duley 1976)

$$T(0, t) - T_0 = \frac{AI(0)w}{K(2\pi)^{1/2}} \tan^{-1} \left[\frac{8\kappa t}{w^2}\right]^{1/2}, \tag{4.8}$$

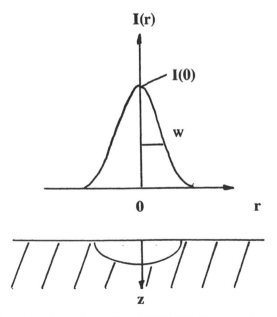

**Figure 4.2.** *Heating of a metal surface with a laser beam with a gaussian intensity distribution.*

where $K$ is the thermal conductivity, $\kappa$ is thermal diffusivity, $w$ is the gaussian beam radius, $T_0$ is the ambient temperature, and $t$ is time. If $T(0,t) = T_m$, the melting temperature, then the laser beam intensity, $I_m(0)$, required to produce melting in time $t$ can be obtained with equation 4.8. $I_m(0)$ is found to be essentially independent of time when $t \gg w^2/8\kappa$ or when the thermal diffusion length $\ell \sim (\kappa t)^{1/2} \gg w$. Because $AI_m(0)$ is obtained as a solution, it is apparent that with $t$ specified, $I_m(0)$ can be reduced through an increase in the absorptivity $A$. The radius of the beam focus on the surface, $w$, will have a profound effect on $I_m(0)$ when $t$ is long but essentially has no effect at short pulse lengths. These solutions do not take into account the latent heat of fusion and thus must be taken to be approximate.

An estimate of the depth of penetration, $z_m$, of the weld pool under spot welding conditions in which melting is included was obtained by Cohen (1967). If $t_m$ is the time at which $T(z = 0) = T_m$, then

$$z_m(t) \sim \frac{0.16\,AI}{\rho L}\,(t - t_m), \tag{4.9}$$

where $\rho$ is the density of the melt and $L$ is the latent heat of fusion. Equation 9 will be strictly valid only when $t_m < 8\kappa/w^2$. With $\rho L = 2 \times 10^3$ J/cm$^3$ and $AI = 10^5$ W/cm$^2$, equation 4.9 predicts $z_m = 8(t - t_m)$ cm or $\sim 10^{-2}$ cm for $(t - t_m) = 1$ msec.

The threshold for conduction welding when the workpiece is moved at a velocity, v, relative to a stationary laser beam delivering an absorbed power, $AP$, to the surface can be obtained from the following approximate solution to the heat equation (Rosenthal 1976)

$$T(r) - T_0 = \frac{AP}{2\pi Kr}\,exp\left[-\left(\frac{v(x + r)}{2\kappa}\right)\right], \tag{4.10}$$

where the coordinate system is as shown in Figure 4.3 and $r = (x^2 + y^2 + z^2)^{1/2}$. Equation 4.10 is the exact solution for a point source of strength $AP$ and has a singularity at $r = 0$. It can be used to estimate the threshold for welding for a gaussian beam of radius, $w$, by taking

$$P = \int_0^\infty I(r)2\pi r dr$$
$$= 2\pi I(0) \int_0^\infty \left(exp\left[\frac{-2r^2}{w^2}\right]\right) r dr \tag{4.11}$$

and by solving for $T = T_m$ at $r = w\sqrt{2}$ and $x = 0$. Then

$$v_m = -\frac{2\sqrt{2}\kappa}{w}\,ln\left[\frac{(T_m - T_0)2\sqrt{2}\pi Kw}{AP}\right], \tag{4.12}$$

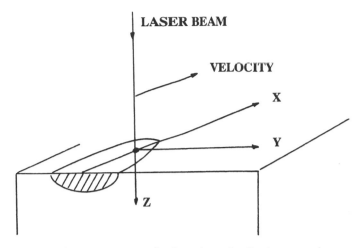

**Figure 4.3.** *Coordinate system for laser beam heating by a scanning source.*

where $v_m$ is the critical speed at which melting will first occur. Numerically, for Al and steel

$$v_m(\text{Al}) = -\frac{2.7 \times 10^{-4}}{w} \ln\left[\frac{2.2 \times 10^7 \, w}{P}\right] \qquad (4.13)$$

$$v_m(\text{steel}) = \frac{-2.5 \times 10^{-5}}{w} \ln\left[\frac{5.7 \times 10^6 \, w}{P}\right] \qquad (4.14)$$

where $P$ is given in watts, $w$ is given in meters, and $v_m$ is given in meters per second. Absorptivities (at 10.6 $\mu$m) have been taken to be A (steel) = 0.1 and Al (Al) = 0.03. Solutions to equations 4.13 and 4.14 are plotted in Figure 4.4. Note that because of the negative sign in the right side of equation 4.12, the allowed solutions for $v_m$ are limited to laser powers in excess of a threshold power $P*$ where:

$$P* = \frac{2\pi K w}{A}(T_m - T_0) \qquad (4.15)$$

This is $2.2 \times 10^7 \, w$ and $5.7 \times 10^6 \, w$ (in watts) for Al and steel, respectively. A more rigorous calculation of the threshold power for conduction welding is given by Cline and Anthony (1977).

## 4.4 THRESHOLD FOR KEYHOLE WELDING

The formation of a keyhole is of fundamental importance for penetration welding. The way in which this occurs is poorly understood in detail but begins with vaporiza-

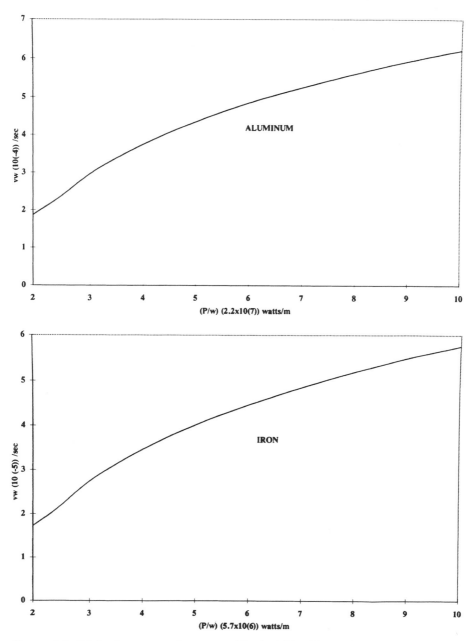

**Figure 4.4.** Solution for $v_m$ versus laser power P for (top) Al and (bottom) steel using the approximate solution given by equation 12. The laser wavelength is 10.6 $\mu m$.

tion at the surface of a weld pool. For a planar surface and an incident beam with a gaussian intensity profile, the solution is given by equation 8 with $T(0,t) = T_v$ the vaporization temperature. Then

$$T_v - T_0 = \frac{A_v I(0)}{K_v (2\pi)^{1/2}} \tan^{-1} \left[ \frac{8\kappa_v t}{w^2} \right], \qquad (4.16)$$

where the subscripts refer to values at or near the vaporization temperature. In many cases, this information is not available, so approximate values must be used.

As for melting, the threshold for vaporization is found to depend on focal radius and pulse duration but becomes independent of time for long pulse lengths; this again assumes that the surface absorptivity is independent of time. The disruption of the surface when vaporization begins will, in general, lead to enhanced coupling and the possibility of a "thermal runaway" effect at the laser focus.

In the initial stage of vaporization, the metal surface will be unperturbed and normal to the direction of incident laser radiation. Under these conditions, a simple one-dimensional vaporization model may be used (Duley 1996). The mass evaporation rate, $\beta$, is related to vapor pressure $p(T)$ as follows:

$$\beta(T) = p(T) \left[ \frac{\overline{m}}{2\pi k T} \right]^{1/2}, \qquad (4.17)$$

where $\overline{m}$ is the average mass of an evaporating atom, $k$ is Boltzmann's constant, and $T(°K)$ is surface temperature. The vapor pressure is given by the Clausius-Clapeyron equation:

$$p(T) = p(T_B) \exp \left\{ \frac{\overline{m} L_v}{\rho k} \left[ \frac{1}{T_B} - \frac{1}{T} \right] \right\}, \qquad (4.18)$$

where $T_B(°K)$ is the boiling temperature and $L_V$ is the latent heat of vaporization (J/$m^3$). The linear vaporization rate is

$$v = \frac{\beta(T)}{\rho}, \qquad (4.19)$$

where $\rho$ is the metal density. At the normal boiling temperature, $v_B = 1.6 \times 10^{-2}$ and $0.73 \times 10^{-2}$ m/sec for Al and Fe, respectively.

For optimal vaporization, the kinetic vaporization rate, $v$, must be equal to that limited by conservation of energy. Then

$$\frac{\beta(\overline{T})}{\rho} = \frac{AI}{L_f + L_v + C(\overline{T} - T_0)}, \qquad (4.20)$$

where $C$ is the heat capacity (J/m³ °K), $T_0$ is ambient temperature, and $L_f$ is the latent heat of fusion (J/m³). The equality given by equation 4.20 defines a temperature $\overline{T}$ that need not be the normal boiling temperature, $T_B$. Detailed calculations show that $\overline{T} \sim T_B$ except at very high intensity ($I > 10^8$ W/cm²) for $CO_2$ laser radiation incident on metals (Duley 1976). At high laser intensity, the linear speed of vaporization, v, approaches the speed of sound in the metal. At the typical laser intensities used in laser welding of metals, $v \sim 1$–$10^2$ cm/sec, and the establishment of the keyhole can occur very rapidly, particularly when geometrical effects lead to enhanced coupling of incident laser radiation once vaporization begins. Theoretical discussions of the vaporization of metals by laser radiation are given by Rykalin, Uglov, and Kokora et al. (1978), Anisimov, Borch-Bruevich, Elyashevich et al. (1967), Finke and Simon (1990), and Metzbower (1993a).

The threshold for the formation of a keyhole also can be affected by surface tension and convective flow in the laser weld pool near the vaporization temperature. Both processes can cause an upwelling of liquid at the periphery of the weld pool (Figure 4.5), which will facilitate self-focussing of incident laser radiation at the center of the weld pool. The enhancement in intensity at this point can be substantial and can assist in the initiation of a keyhole. Experimental evidence for the development of a concave liquid surface at the threshold for formation of a keyhole was reported by Mazumder and Voekel (1992) (Figure 4.6).

An intrinsic instability in laser-irradiated surfaces also may be responsible for enhanced coupling at high laser intensity (Ursu et al. 1987, Jain et al. 1981). The effect involves the production of a grating-like structure with periodicities comparable to the laser wavelength (Isenor 1977, Keilmann and Bai 1982, Young et al. 1983). This is accompanied by enhanced absorption of laser radiation, with coupling efficiencies approaching unity (Ursu et al. 1984). It has been suggested that this increased coupling is responsible for the large drop in metal reflectivity seen near the intensity threshold for laser welding (Keilmann 1983).

**LASER BEAM**

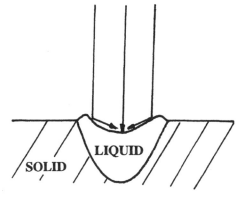

*Figure 4.5.* Schematic representation of weld pool morphology due to convection and surface tension. This causes self-focussing of an incident laser beam.

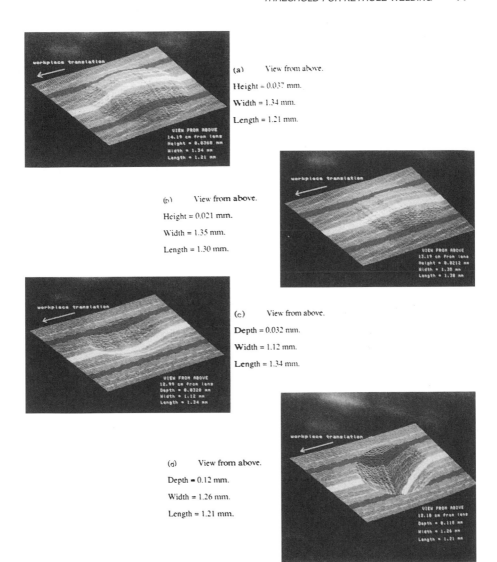

**Figure 4.6.** *Shape of weld pool surface as measured by Mazumder and Voekel (1992). Incident laser intensity increases from (a) to (d).*

These effects, which enhance the absorption of laser radiation at the surface of the weld pool, are responsible for a sharp increase in laser coupling at the threshold for keyhole formation. This is accompanied by a dramatic increase in weld penetration (Figure 4.7). Data given in Figure 4.7 show that the effect is most noticeable at the $CO_2$ wavelength. This arises from the initial low absorptivity at 10.6 $\mu$m.

At the onset of vaporization, gas is ejected from the region on the surface of the weld pool at which the laser intensity is highest. This is expected to be close to the

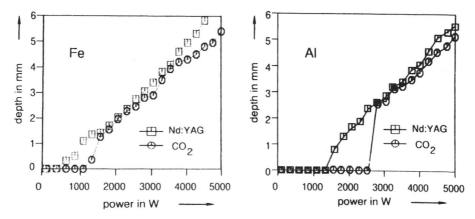

**Figure 4.7.** *Welding depth in Fe and Al versus incident laser power. The laser wavelengths are 10.6 $\mu$m ($CO_2$) and 1.06 $\mu$m (YAG). From Hügel et al. (1994).*

center of the weld pool in the case of a gaussian or quasi-gaussian laser intensity distribution (Metzbower 1993). The liquid is then subjected to a recoil pressure equal to the vapor pressure:

$$p(T) \; = \; \beta(T) \left[\frac{2\pi kT}{m}\right]^{1/2} \tag{4.21}$$

and will flow from the center of the weld pool out toward the periphery, enhancing the concavity of the weld pool surface. This will have the effect of increasing the intensity of laser radiation at the center promoting vaporization. The concave structure generated in this way gradually deepens and assumes a higher aspect ratio. In fact, it appears that this may occur in two stages. In the first stage, a shallow concavity is formed with the depth at the center comparable to the radius of the weld pool. Self-focussing by this structure concentrates the laser radiation at the center of this liquid concavity, where another, higher aspect, ratio hole, which may be more cylindrical, is formed. It is this extended high aspect ratio structure that is the primary form of the keyhole that leads to efficient coupling of laser radiation into the workpiece and to penetration welding conditions.

In the absence of motion of the workpiece or beam, the keyhole created in this way would assume the geometry commonly observed during laser drilling of metals (von Allmen 1976, Chryssolouris 1991). This geometry is a complex function of beam focussing conditions, material properties, and time. For example, it is not unusual to have the drilled hole first increase in diameter with depth, before narrowing to a smaller diameter farther inside the material. For simplicity, however, we will consider a conical structure (Figure 4.8) with a cone angle $2\bar{\theta}_w$. This structure is an efficient trap for incident laser radiation. According to Kaplan (1994), the beam intensity after n reflections is

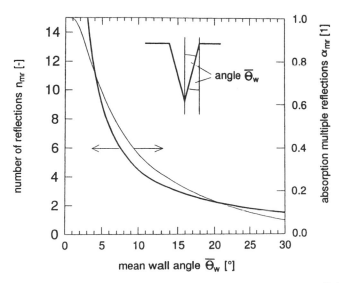

**Figure 4.8.** *Number of reflections (thick line) and corresponding absorption coefficient (thin line) for a cone with wall angle $\bar{\theta}_w$. R is assumed to be 0.88. From Kaplan (1994).*

$$\frac{I(n)}{I(0)} = R^n, \tag{4.22}$$

where $R$ is the Fresnel reflection coefficient at the angle of incidence. The maximum number of reflections is

$$n_m = \frac{\pi}{4\overline{\theta}_w}, \tag{4.23}$$

which need not be an integer number. Multiple reflections rapidly damp the laser intensity and result in heat input through the walls of the keyhole (Wang and Wei 1992).

If $r_k$ is the radius of the keyhole and $d$ is its depth, then $n_m$ becomes

$$n_m = \frac{\pi}{4 \tan^{-1}(r_k/d)} \tag{4.24}$$

and the absorptivity is approximately

$$A = [1 - R^{n_m}]. \tag{4.25}$$

This result does not include the effect of plasma absorption within the keyhole (Miyamoto et al. 1986, Kaplan 1994). If one allows for absorption and scattering of incident laser radiation in the gas over the keyhole, then an approximate efficiency factor for energy deposition in the keyhole, $\eta_k$, can be defined (Fuerschbach and MacCallum 1995) as follows:

$$\eta_k \sim (1 - A_p)(1 - R^{n_m}). \tag{4.26}$$

As a typical value one has $\eta_k = 0.5$ when $1 - A_p = 0.9$, $R = 0.9$, $r_k = 0.1$ mm, and $d = 1$ mm.

This result is derived on the basis of geometrical optics, and will be valid when the diameter of the keyhole $2r \gg \lambda$. When the keyhole tapers to a smaller diameter, diffraction will become important and the beam will spread as the result of diffraction (Duley 1987). The angle, $\theta$, that a parallel beam will spread on passing through an aperture of diameter $2r$ is

$$\theta \sim \sin^{-1}\left[\frac{\lambda}{2r}\right]. \tag{4.27}$$

The limiting diameter for the keyhole then corresponds to $2r \sim \lambda$ where $\theta = \pi/2$. Equation 4.27 may also explain why the keyhole seems to be initiated from an area with a radius $r_k \ll w$ the laser beam radius near the center of the weld pool (Figure 4.9). Rays reflected from the periphery of the laser-induced melt pool arrive at a large angle $\theta$ with respect to the normal to the surface of the weld pool. When $\theta$ approaches an angle of 90°, coupling occurs efficiently into an aperture of radius $r_k = \lambda/2$, raising the intensity at this point and enhancing vaporization.

The propagation of infrared radiation in hollow cylindrical metallic waveguides has been discussed by Marcatili and Schmeltzer (1964), Garmire, McMahon, and

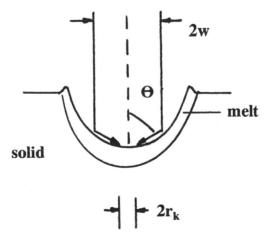

**Figure 4.9.** Diffraction at angle $\theta$ is responsible for initial coupling of reflected beams into a small keyhole.

Bass (1976), and Gu and Morrow (1994). The attenuation factor for a waveguide of radius $r$ is

$$\alpha = \frac{\gamma \lambda^2}{r^3},$$ (4.28)

where $\gamma$ is a term that depends on the propagation mode and material properties but is typically $\sim 10^{-2}$ for metals. Then, with $r = \lambda$, $\alpha \cong 10^3$ m$^{-1}$ or 10 cm$^{-1}$. This indicates that incident laser power will be attenuated over a depth of $\sim 1$ mm in such a structure. The formation of a waveguide-like cylinder with a radius of $\sim \lambda$ may therefore be the initial step in the creation of the keyhole involved in penetration welding of metals. This structure will be stable only on a transient basis because of its small diameter.

For anything other than normal incidence, the coupling of laser radiation into a metal will be dependent on the beam polarization. The reflectivity for the polarization vector in the plane of the metal surface (s-polarization) remains high at all angles of incidence, whereas that for the polarization vector in the plane of incidence (p-polarization) can become very small at oblique angles (Born and Wolf 1975) (Figure 4.10). As a result, once a keyhole has formed, coupling of incident laser radiation will be polarization dependent. This effect has been used to optimize laser welding conditions (Arzuov et al. 1979, Olsen 1980, Garashchuk, Kirsei, and Skinkarev 1986, Beyer, Behler, and Herziger 1988, Sato, Takahashi, and Mehmetli 1996).

The keyhole is initially formed through drilling of a hole. This drilling phase occurs on a time scale that is short compared with that for lateral heat conduction

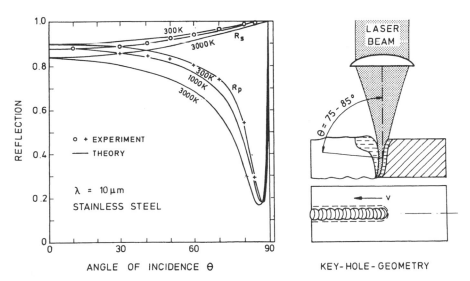

**Figure 4.10.** *Reflectivities, $R_s$ and $R_p$ at 10.6 $\mu$m for stainless steel at various temperatures. From Behler et al. (1988a).*

through the walls of the keyhole into the surrounding medium. This is apparent from the geometry shown in Figure 4.8 because, with an incident intensity

$$I(r) = I(0)\exp - [r^2/w^2];$$    (4.29)

the intensity is always largest on the axis of beam propagation. In addition, because the walls of the keyhole assume an angle $\bar{\theta}_w$, the intensity absorbed at the walls will be substantially less than that absorbed at the center of the beam and will decrease with time to reach a limiting value that is below the threshold for fast vaporization. The velocity of expansion of the fusion front in a lateral direction from the wall of the keyhole is approximately

$$v_m = \frac{A(r)I(r)}{\Delta H_m},$$    (4.30)

where $A(r)$ is the absorptivity at position $r$, $I(r)$ is as given by equation 4.29, and $\Delta H_m$ is the latent heat of fusion. The drilling velocity in the direction of beam propagation is approximately

$$v_V = \frac{A(0)I(0)}{\Delta H_V},$$    (4.31)

where $A(0)$ is the absorptivity at the center of the beam. The ratio

$$\frac{V_m}{V_v} = \frac{A(r)}{A(0)} \frac{I(r)}{I(0)} \frac{\Delta H_V}{\Delta H_m}$$    (4.32)

$$= \frac{A(r)}{A(0)} \frac{\Delta H_V}{\Delta H_m} \exp(-r^2/w^2)$$    (4.33)

is then an approximate measure of the lateral fusion depth to the depth of the keyhole at short times, t, after the onset of keyhole formation. For a metal such as iron, $\Delta H_V \sim \Delta H_m$, but this is offset by the small absorptivity at r due to the high inclination angle of the keyhole wall. Multiple reflections in the keyhole also will tend to increase I(0). The overall result is to introduce a time delay, $\Delta t$, between drilling of the keyhole and the formation of an appreciable liquid sheath around the keyhole. This time delay is approximately $\Delta t(r) = r/v_m$ or the time it takes the fusion front to propagate a distance comparable to the radius of the keyhole. After the time $\Delta t$, the keyhole is surrounded by a liquid sheath whose volume is comparable to that of the keyhole. At this point in time, instabilities in the keyhole liquid interface will start to dominate the interaction. These instabilities would cause the depth of the keyhole to fluctuate even in the absence of motion of the workpiece.

The time delay $\Delta t$ between establishment of the keyhole and lateral heat transfer from its walls can be seen in simulations (Olfert 1998). In this simulation, the development of a drill hole in solid paraffin exposed to 10.6-$\mu$m $CO_2$ laser radiation was imaged at 30 Hz and show that the maximum depth of the keyhole is rapidly established, whereas lateral heat transfer is seen only after the maximum penetration depth is attained. Once the volume of liquid surrounding the drill hole is comparable to that of the hole itself, instabilities develop that cause the drill hole to be periodically interrupted. When this occurs, liquid also moves to the tip of the drill hole filling up part of its volume and reducing the penetration depth. The overall effect of this accumulation of liquid and instabilities in the perimeter of the keyhole is to decrease the average penetration depth, although occasional spiking can reestablish the initial depth of the drill hole. This spiking arises from intermittent closure of the channel leading to the end of the drill hole due to liquid flow and occurs even at incident laser intensities well below the threshold for plasma initiation. Spiking at the end of the keyhole during laser welding is attributable to this effect as well as to the interruption of the incident laser beam by the plasma generated over the keyhole.

## 4.5  STABILITY AND DYNAMICS OF THE KEYHOLE

The keyhole is basically an unstable structure even in the absence of motion. As demonstrated in Section 4.4, the creation of a liquid sheath around the keyhole due to lateral heat transfer immediately introduces instabilities. To understand how these instabilities arise, we must consider the forces that act on this liquid and how they are generated. As an initial simplification, we will disregard the motion of the keyhole through the workpiece and evaluate the terms that act on a stationary keyhole. A blind hole also will be assumed. A number of theoretical studies of the keyhole problem have been published (Andrews and Atthey 1976, Klemens 1976, Dowden et al. 1987, Beck, Berger, and Hugel 1992, Matsuhiro, Inaba, and Ohji 1992, Kroos, Gratzke, and Simon 1993a, Kroos et al. 1993b, Mueller 1994a, Kaplan 1994, Schulz et al. 1996).

The keyhole, defined in this way, has the simplified geometry shown in Figure 4.11 and in equilibrium is maintained through a balance among the following pressure terms:

$p_g$ = hydrostatic pressure

$p_\sigma$ = surface tension

$p_v$ = vaporization pressure

$p_h$ = hydrodynamic pressure

$p_\ell$ = radiation pressure.

The pressure terms $p_v$, $p_h$, and $p_\ell$ tend to keep the keyhole open, whereas $p_g$ and $p_\sigma$ are restoring pressures. All terms depend on depth $z$ and keyhole radius $r(z)$. In equilibrium,

$$p_v + p_l = p_\sigma + p_g + p_h, \tag{4.34}$$

at all points $(r, z)$ on the surface of the keyhole. The functional form for these pressure terms is summarized in Table 4.2, and typical values are given under laser

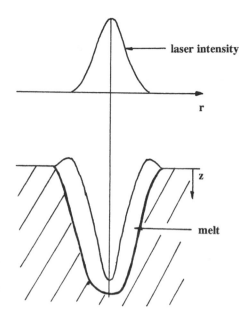

**Figure 4.11.** *Simplified geometry for stationary keyhole.*

welding conditions for metals. It is apparent that $p_\ell$ and $p_g$ are small compared with such terms as $p_\sigma$ and $p_v$, and we do not consider these terms in the following discussion. The hydrodynamic term $p_h$ also is small at low welding speed and can be neglected. Then

$$p_\sigma \sim p_v \qquad (4.35)$$

**TABLE 4.2. Pressure Terms in Laser Keyhole**

| Pressure Term | Functional Form | Typical Value (N/m²) | (atm) |
|---|---|---|---|
| $p_g$ | $\rho_\ell g z$ | 75 | $7.5 \times 10^{-4}$ |
| $p_\ell$ | $\dfrac{I}{c}[A + 2R]$ | 50 | $5 \times 10^{-4}$ |
| $p_\sigma$ | $\sigma\left[\dfrac{1}{r_1(z)} + \dfrac{1}{r_2(z)}\right]$ | $10^4$ | $10^{-1}$ |
| $p_h$ | $\dfrac{\rho_\ell}{2}[v_{min}^2(z) - v_{max}^2(z)]$ | $0^*$ $10^2-10^4$ | $0^*$ $10^{-3}-10^{-1}$ |
| $p_v$ | $-\overline{m}n_g u_g^2$ | $10^4$ | $10^{-1}$ |

Numerical values are for a metal such as Fe.

* For low welding speed.

$p_\ell$ = liquid density (kg/m³); $g$ = acceleration due to gravity (m/sec²); $z$ = depth of keyhole; $I$ = laser intensity (W/m²); $c$ = speed of light (m/sec); $A$ = absorptivity; $R$ = reflectivity; $\sigma$ = surface tension (N/m); $r_1$, $r_2$ = keyhole radii (m); $v_{min}$, $v_{max}$ = liquid flow velocity around keyhole (m/sec); $\overline{m}$ = mass of vaporizing atom (kg); $n_g$ = density at surface of Knudsen layer (m⁻³); $u_g$ = gas velocity at Knudsen layer (m/sec).

is the defining equation for pressure balance and determines the stability of the keyhole. With $p_v \sim 0.1$ atm ($10^4$ N/m$^2$) and using $\sigma \sim 1$ N/m, equation 4.35 is satisfied with $r_1 \sim 10^{-4}$ m; $r_2$ is the radius of the keyhole surface in the direction of laser propagation and so will be large ($r_2 \gg r_1$), except at the tip of the keyhole. Because $r_1$ also is small at this point, the pressure is highest at the tip of the keyhole and diminishes along its length to the entrance hole.

Calculation of the ablation pressure is complicated by the formation of a Knudsen layer over the vaporizing surface. This layer will have a thickness, $\delta$, which is several mean free paths. With

$$\delta = \frac{1}{\sigma_c n}, \tag{4.36}$$

where $\sigma_c$ is the collision cross section and n is the atomic density in the gas. Taking $\sigma_c = 10^{-16}$ cm$^2$ and n $\sim 3 \times 10^{19}$ cm$^{-3}$, $\delta = 3 \times 10^{-4}$ cm. This is smaller than the diameter of the keyhole over much of its length but comparable to the diameter at the tip of a closed keyhole.

The Knudsen layer has the effect of redistributing particle velocity from that range emitted by the vaporizing surface to a Maxwellian distribution. Redeposition of particles may occur at the surface effectively reducing the vaporization rate. The pressure exerted on the vaporizing surface in the presence of a Knudsen layer is

$$p_v = \overline{m}(n_g u_g^2 - n_l u_l^2) \tag{4.37}$$

where $n_g$, $n_l$ are particle densities at the top surface of the Knudsen layer and over the liquid surface, respectively; $n_g$, $n_l$ are corresponding particle velocities. Because *mnu* is a constant (i.e., no particles are created or destroyed in the Knudsen layer), with $n_l > n_g$, one obtains the approximate result given in Table 4.2, namely,

$$p_v \sim \overline{m} n_g u_g^2 \tag{4.38}$$

$p_v$ can be evaluated when the surface temperature, $T_s$, is known. Self-consistent solutions relating keyhole radius, $T_s$, and $p_v$ were obtained numerically by Kroos, Gratzke, and Simon (1993a). The result is the relationship among $p_v$, $p_\tau$, and keyhole radius, a, as shown schematically in Figure 4.12 and calculated for a cylindrical keyhole and finite scan speed. This plot shows that the relative excess pressure, ($p + p_0)/p_0$, where $p_0$ is the ambient pressure, arising from the vaporation and surface tension terms have two solutions for $a/r_0$ where $p_v = p_\tau$. One of these occurs at small $a/r_0$ values, whereas another is found at a much larger keyhole radius. The vaporization pressure, $p_v$, first increases as a increases as the larger keyhole radius allows more of the incident laser beam to enter the keyhole. Due to the assumed Gaussian beam profile, $p_v$ decreases when a $>1.5$ $r_0$ because the laser beam can then pass through the keyhole without heating the walls.

The solution to the left of point N in Figure 4.12 is unstable because $p_\sigma > p_v$ and the keyhole will collapse. At point N, $p_\sigma = p_v$, but this represents an unstable equilibrium because an increase in a yields an increase in $p_v$ while $p_\sigma$ decreases. Thus, for points to the right of N, the keyhole will continue to expand. The stable

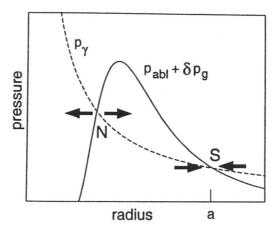

**Figure 4.12.** Schematic representation of the dependence of ablation and surface tension pressure on keyhole radius, a. From Kroos, Gratzke, and Simon (1993a).

solution occurs only at point S because a change in a near this point will result in a restoring force. One expects, therefore, that the keyhole radius would stabilize at $a_B$ and that oscillatory solutions would be possible about this radius.

Kroos, Gratzke, and Simon (1993a) found that the range $a_S-a_N$ depends on the ratio $P/d$, where $d$ is the depth of the uniform cylindrical keyhole, and that no solution is possible for $P/d$ of less than a minimum value. It was suggested that this value is associated with the threshold intensity for the development of a keyhole. As $P/d$ approaches this minimum value, the solution becomes increasingly unstable as $a_S-a_N \rightarrow 0$. In this range, the penetration depth would fluctuate.

These idealized solutions also can be used to explore the effect of noncylindrical keyhole geometries. A reduction in P/d results in the stable solution moving to smaller a. By replacing d by an incremental length $\Delta z$ where $z$ is the depth coordinate (Figure 4.11) and letting

$$P(z) = P(0)\exp(-\alpha z) \tag{4.39}$$

where $\alpha$ is the effective absorption coefficient for laser radiation within the keyhole,

$$\frac{P}{d} \equiv \frac{P(z)}{\Delta z} = \frac{P(0)}{\Delta z} \exp(-\alpha z) \tag{4.40}$$

is obtained as the incremental absorbed power per unit length at depth $z$ within the keyhole. With $\Delta z$ constant, it is apparent that $P/d$ decreases as $z$ increases. The solution for $a/r_0$ then moves to smaller values until the limiting value of $P(z)/\Delta z$ is reached. This defines the maximum depth of the keyhole. The overall effect is to produce a keyhole whose radius tapers from a large value at $z = 0$ to $a \sim r_0$ at the maximum penetration depth. This is accompanied by a gradient in the vapor pressure $p_v$ such that $p_v$ is greatest at the end of the keyhole.

This model, although idealized, reproduces many of the properties of the keyhole obtained during laser welding at slow speed and can be used to estimate the dynamic effect of perturbations on the solution.

A cavity such as the keyhole surrounded by liquid will have a range of characteristic mechanical vibrational modes (eigenmodes) that can support oscillation at a variety of frequencies (eigenfrequencies). Identification of these modes in the case of a cylindrical keyhole is straightforward (Figure 4.13). Excitation and damping of these modes will depend on such factors as gas flow in the keyhole, its diameter and length, the volume of the liquid surrounding the keyhole, mechanical and fluid properties of this liquid, instabilities in laser intensity, and boundary conditions at the input and exit of the keyhole. Solutions to this problem have been obtained under a range of simplifying assumptions by Postacioglu et al. (1987), Postacioglu, Kapadia, and Dowden (1989, 1991), Kroos et al. (1993a,b), Ledenev, Mirzoyer, and Nikolo (1994), Banishev et al. (1994), and Klein et al. (1994), and Klein, Vicanek, and Simon (1996). A range of frequencies should be possible as various azimuthal and axial modes are excited. The simplest of these modes is the pure radial vibration with mode numbers $(n, \ell) = (0, 0)$ as shown in Figure 4.13. According to Klein et al. (1994), the time-dependent radius is

$$r_s(t) = a + \alpha(t), \qquad (4.41)$$

where $\alpha \ll a$ is a requirement for linearity. Excitation of this mode will be due to a balance between the dynamic pressure within the keyhole and that due to surface

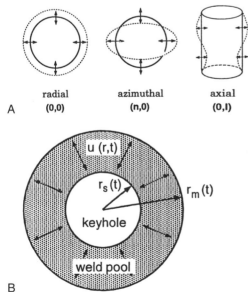

radial     azimuthal     axial
A    (0,0)       (n,0)        (0,l)

B

**Figure 4.13.** Top: basic eigenmodes for oscillation in a cylindrical cavity. Azimuthal and axial mode numbers are n and ℓ, respectively. From Klein et al. (1994). Bottom: cross section of weld pool transverse to laser beam.

tension. The eigenfrequency $w_{00}$ for the motion of the melt in the (0, 0) mode is shown by Klein et al. (1994) to be

$$w_{00} \cong \left[ \frac{B}{\rho a \ell n C} \right]^{1/2} \tag{4.42}$$

where

$$B = \frac{\partial}{\partial r} [p_\sigma - p_v - \delta p_g]_{r=a} \tag{4.43}$$

and $C = a_m/a$, where $a_m$ is the outer radius of the weld pool (Figure 4.13b). $\delta p_g$ is the excess pressure within the keyhole due to gas flow, and

$$\delta p_g \sim \frac{\overline{m}}{3} n_g u_g^2 \left[ \frac{d}{a} \right]^2, \tag{4.44}$$

where d is the length of the cylindrical keyhole.

For metals such as Fe, Al, and Cu, $w_{00} \sim 1.5$–$2$ kHz for $d = 1$ mm and typical welding conditions (Klein et al. 1994). The calculated dependence of $w_{00}$ on laser power and welding speed is shown in Figure 4.14.

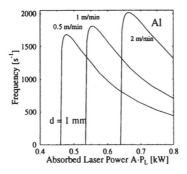
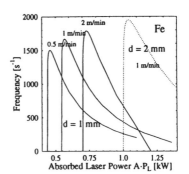
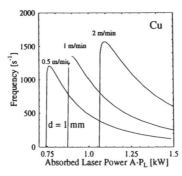

**Figure 4.14.** *Calculated fundamental mode frequencies (0, 0) for Al, Fe, and Cu and various sheet thicknesses. From Klein et al. (1994).*

Klein et al. (1994) also derived an expression for the eigenfrequencies $w_{n\ell}$ associated with eigenvibrations $(n, \ell)$. This is

$$w_n\ell = \left\{ \frac{ka|K'_n(ka)|}{|K_n(ka)|} \left[ \frac{\sigma}{\rho a^3} (n^2 + k^2a^2) + \frac{B}{\rho a} \right] \right\}^{1/2} \tag{4.45}$$

where $k = 2\pi/\lambda$ is the wavenumber of the axial mode, and $\lambda$ is its wavelength, $K_n(ka)$ is a modified Bessel function and $K'_n(ka)$ is the derivative with respect to its argument, and $B$ is the restoring pressure constant (equation 4.42). The frequencies, $w_{n\ell}$ lie in the range of 2–8 kHz for Fe d = 1 mm and form bands of allowed energy states (Figure 4.15). These solutions become degenerate in $\ell$ as d becomes large.

Damping of these modes arises due to dynamic friction and is characterized by a lifetime $\Gamma^{-1}$ where

$$\Gamma_{oo} = \frac{\eta}{\rho a^2 \ln c}, \tag{4.46}$$

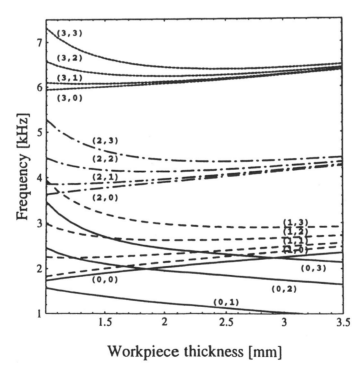

**Figure 4.15.** *Maximum eigenfrequencies of radial, axial, and azimuthal eigenmodes for Fe plate. The laser beam radius is 150, μm. From Klein et al. (1994).*

where $\eta$ is the dynamic viscosity of the liquid surrounding the keyhole. Low-order modes are found to have lifetimes in the range of 50–100 msec, whereas this reduces to 1–10 msec for modes with $(n, \ell) > (3, 3)$. Because the oscillation frequencies are in the range of 2–8 kHz, modes are damped over ~50 oscillations for (0, 0) and ~10 oscillations for the higher-frequency vibrations.

The existence of a range of resonant vibrations in the keyholeliquid sheath layer dominates the perturbation response of the system. Perturbations can arise through a variety of terms, including fluctuations in laser power, gas pressure, mechanical vibrations, and so on. The perturbation response was studied on a theoretical basis by Klein, Vicanek, and Simon (1996) and on an experimental basis by Gu and Duley (1991) and Heyn, Decker, and Wohlfahrt (1995). The simplest approach is to introduce a periodic term in the incident laser intensity, through a modulation in laser power. An example of the effect of beam modulation on the spectrum of acoustic emission during $CO_2$ laser welding of mild steel in the bead on plate configuration is shown in Figure 4.16. The sharp resonances that appear in the acoustic spectrum at high modulation depth correspond to multiples of the modulation frequency and indicate a nonlinear response. Klein, Vicanek, and Simon (1996) concluded that laser power fluctuations as small as 1% at the resonant frequency may be sufficient to cause keyhole collapse. This response can be invoked as long as a multiple of the modulation frequency, f, lies within one of the frequency bands corresponding to natural modes of the system (Figure 4.15). Thus, a strong response is not necessarily produced at the modulation frequency, but instead "amplified" components are apparent when $Nf \sim w(n, \ell)$ where $N$ is some integer. The envelope connecting the emission peaks at $Nf$ contains more than one maximum because these reflect the density of states function for acoustic resonances within the system.

Similar discrete frequency terms at $Nf$ are observed in the optical emission detected during welding with modulated beams (Gu and Duley 1996). This suggests that the acoustic modes couple directly into the excitation of plasma within and in front of the keyhole (i.e., that the pressure term $\delta p_g$ [equation 4.44] is time-dependent with frequency components at $Nf$). More generally, in the absence of forced oscillation, $\delta p_g$ will reflect the noise spectrum of the system and lead to a feedback term that causes certain noise frequencies to be amplified. Fast Fourier transform spectra of acoustic and optical emission during laser welding (Figure 4.16) reflect this amplified noise spectrum and show a variety of sharp peaks superimposed onto a background of many overlapped frequency components. A broad spectral component often observed near 4.5 kHz can be associated with the fundamental resonance of the keyhole vibration spectrum.

Competition between a driven and spontaneous response occurs naturally in such systems when the laser intensity contains a modulation term. This offers the possibility of stabilization of the keyhole through excitation at specific frequencies. Such stabilization might be expected to lead to more uniform welding conditions (i.e., to uniform penetration depth and freedom from weld defects).

A partial simulation of these keyhole effects is possible through a study of the interaction between focussed $CO_2$ laser radiation and an absorbing liquid (Duley et al. 1992, Olfert et al. 1994, Olfert and Duley 1996). Imaging of the keyhole shows

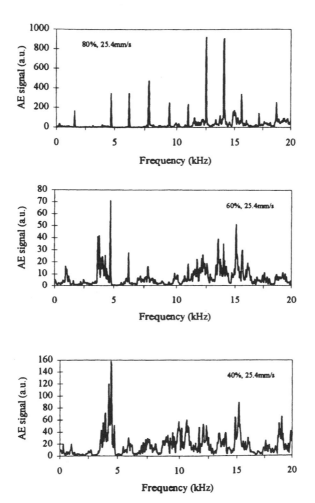

**Figure 4.16.** *Acoustic spectra obtained during welding of mild steel at different modulation depths with a 50% duty cycle. The laser intensity was 1.1 × 10^6 W/cm². From Gu and Duley (1996a).*

how keyhole morphology evolves with time and is influenced by beam modulation. Some data are given in Figure 4.17; a continuous-wave $CO_2$ laser beam was chopped mechanically and focussed on the surface of liquid glycerol. The resulting morphology of the keyhole was recorded with a CCD camera at 30 Hz and then averaged. Edge-enhanced images of the keyhole plotted in Figure 4.17 for various modulation frequencies show how the structure of the keyhole evolves from a deep penetration mode at $f = 2760$ Hz, exhibiting large fluctuations in penetration with time, to a stabilized structure at low frequency. This stable structure consists of a broad hemispherical depression with a radius of ~0.5 times the maximum keyhole depth and centered on the entrance of the keyhole. A narrow quasicylindrical extension

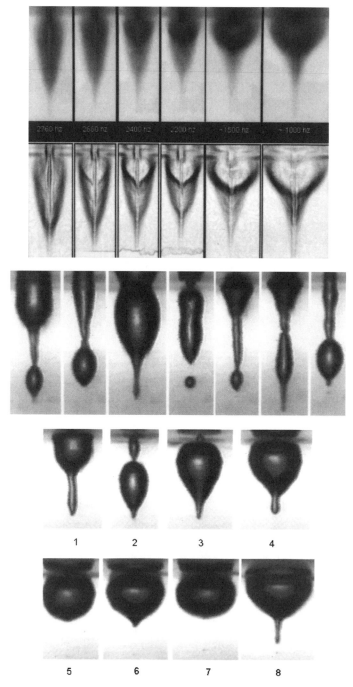

**Figure 4.17.** *(Top) Frame-averaged images of $CO_2$ laser drilling of liquid glycerol with beam modulation at varying f values. (Middle) typical frame sequence at 2750 Hz. (Bottom) frame sequence at 1300 Hz.*

to the keyhole begins about half-way down the keyhole. Frame-to-frame images recorded at 2750 and 1300 Hz (Figure 4.17) show this transition from a basically unstable, random interaction at high modulation frequency to the resonant mode at low frequencies. A resonance can be achieved over a range of modulation frequencies, reflecting the band-like nature of the eigenstates of this system.

Interruption of the incident laser beam by collapse of the keyhole is responsible for the disrupted structure seen in Figure 4.17, middle. When this occurs, liquid from the surrounding volume is momentarily in a region of high incident laser intensity. This results in rapid heating and vaporization and the generation of gas that expands in both directions along the beam propagation axis. Gas moving back toward the direction of the laser can be vented through the entrance of the keyhole; however, gas moving toward the blind end of the keyhole accumulates to form a bubble that expands laterally. The presence of this bubble allows laser radiation to penetrate further into the material, so a momentary spiking occurs in the penetration depth. As a result, the keyhole in Figure 4.17, middle, has the appearance of a number of open structures connected by "hourglass" constrictions.

At modulation frequencies within the range of system eigenfrequencies (Figure 4.17c), keyhole closure occurs at or near the surface, and the interaction presumably is stabilized through the presence of the surface, such that the bubble produced as the result of vaporization at the point of closure maintains a constant volume. Occasionally, laser radiation penetrates through the bubble to impinge on its inner surface; this produces the quasicylindrical structure seen at the bottom of the bubble in Figure 4.17, bottom, 1–4, 6, and 8.

The periodic nature of the system response during continuous-wave excitation in the absence of beam modulation is evident in the image shown in Figure 4.18

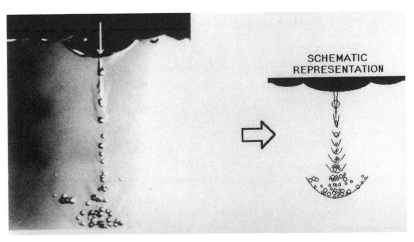

**Figure 4.18.** *Video frame taken ~1.3 seconds into zero g drill into 66% glycerol (33% water) at $I_0 = 0.9 \times 10^5$ W/cm². Hemispherical striations propagating away from the end of the drill hole associated with groups of ejected bubbles are clearly visible, indicating a periodic driving mechanism. From Olfert et al. (1994).*

obtained under zero gravity conditions (Olfert et al. 1994). Vaporization at a point close to the liquid surface is seen to yield gas bubbles that are driven into the liquid. These accumulate at a pseudocontact front structure at a depth many times that of the keyhole. The formation of each bubble transiently opens up the keyhole, allowing laser radiation to penetrate to a greater depth.

Although the simulation of keyhole formation and structure shown in Figure 4.17 is of limited usefulness in describing the behavior of keyholes formed during the welding of metals, some general similarities exist. These can be summarized as follows:

1. Any motion of the walls of the keyhole changes the interaction with the incident beam and thus results in a pressure variation within the keyhole.
2. Vaporization of liquid moving into the keyhole will be rapid and will lead to gas flow in both directions along the beam. Gas driven toward the tip of the keyhole will produce bubbles that may remain trapped after solidification.
3. If this pressure pulse is synchronized with a driving term in the laser intensity through beam modulation, resonant stabilization of the keyhole geometry can be produced.
4. Small instabilities in system parameters can grow into large-scale perturbations if they contain frequency components that lie within the range of allowed eigenfrequencies.

## 4.6  MOVING KEYHOLE

The previous discussion is based on an assumption that the keyhole has rotational symmetry about the laser propagation axis. This assumption may be correct at slow welding speeds but does not hold at the high speeds encountered in many laser welding applications. In this case, the leading and trailing edges of the keyhole generally assume different shapes, with the leading edge curving back toward the weldment. Under such conditions, the tip of the keyhole may be displaced significantly from the direction of beam delivery at the top of the keyhole. This geometry has been directly observed in radiographic images during $CO_2$ laser welding of metals and by simulations of laser welding carried out in transparent materials such as glass (Arata 1987, Siekman and Morijn 1968, Duley et al. 1992). Some images of the keyhole are shown during penetration welding of polypropylene with $CO_2$ laser radiation (Figure 4.19). These images show bending of the leading edge back toward the weld as well as the presence of wave-like structures on the liquid surface at the trailing edge. These instabilities occasionally result in momentary closing of the keyhole. It also is apparent that no direct optical path exists between the center of the entrance to the keyhole and the curved tip of the blind end. This is strong evidence for channeling of incident laser radiation via reflection from the leading edge of the keyhole. The absorption of incident laser radiation in such a structure

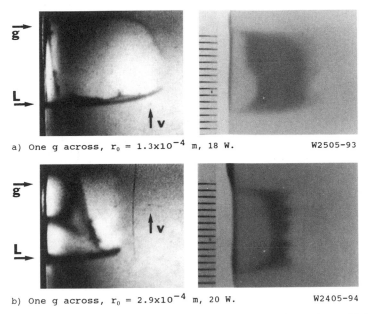

a) One g across, $r_0 = 1.3 \times 10^{-4}$ m, 18 W.          W2505-93

b) One g across, $r_0 = 2.9 \times 10^{-4}$ m, 20 W.          W2405-94

**Figure 4.19.** *Morphology of laser-induced keyhole in welding of polypropylene with $CO_2$ laser radiation. From Mueller (1994b).*

is discussed by Ducharme, Kapadia, and Dowden (1992), Schuöcker and Kaplan (1994), and Kaplan (1994).

Energy transfer in this case must involve Fresnel absorption at the leading edge of the keyhole together with heat transfer from the plasma within the keyhole. Some reflections must also cause laser radiation to impinge on the trailing edge of the keyhole. A simulation of these effects was carried out by Kaplan (1994). Figure 4.20 shows the calculated shape of the keyhole in a plane containing the welding direction and its relation to the melt distribution for $CO_2$ laser welding of steel at 50 mm/sec at a power of 4 kW. Tilting of the tip of the keyhole back away from the beam axis is clearly evident in this simulation. The plasma absorption coefficient $\alpha$ also is shown, and it reaches a maximum value at points slightly off the surface of the leading and trailing edges of the keyhole. Calculated values for $\alpha$ are similar to those measured within the laser keyhole during welding by Miyamoto et al. (1986).

It is apparent that the keyhole shown in Figure 4.20 would revert to a symmetrical shape should the weld speed drop to zero. The leading edge curvature, which produces a trailing tail structure, is maintained as the system attempts to achieve an equilibrium depth for the keyhole. An important parameter, as pointed out by Semak et al. (1994a) and Matsunawa et al. (1996), is the ratio $R = a/vt_d$, where $a$ is the keyhole radius, $v$ is the welding speed, and $t_d$ is the time to drill a hole equal to the maximum penetration depth for a stationary beam. With a $\sim 150$ $\mu$m and $t_d \sim 10$ msec, $R \sim 0.15/v$, where $v$ is given in meters per second. High leading edge curvature and a tendency for weld instabilities to develop are expected when $R$ is $\leq 1$ or v is

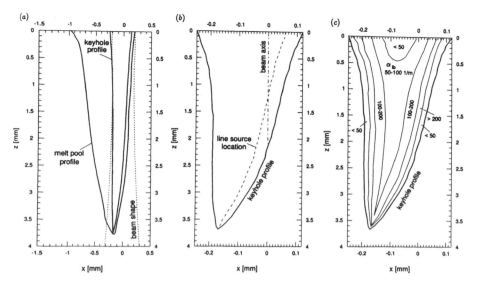

**Figure 4.20.** Shape of beam, keyhole, and molten pool (a), keyhole profile magnified in the x direction (b), and distribution of plasma absorption coefficient $\alpha_{iB}$ in the magnified keyhole (c) for 4 kW, 50 mm/sec, and steel. From Kaplan (1994).

$\geq 1.5$ cm/sec. These instabilities will arise as the leading edge of the keyhole wall alternatively recedes from and catches up with the welding speed. Matsunawa et al. (1996) derived the following expression for the velocity, $v_d$, of the keyhole wall in a direction normal to its surface:

$$v_d = \left[ \frac{\kappa}{w} \frac{\rho_\ell}{\rho_s} \left( \frac{2u_g}{L_v \rho_\ell} \right) \left[ I_a - \frac{K v}{\kappa} (T_V - T_m) \right] - \frac{\sigma}{w} \right]^{1/2} \tag{4.47}$$

where $\kappa$ is thermal diffusivity, $u_g$ is the gas velocity at the Knudsen layer, $L_v$ is the latent heat of vaporization, $I_a$ is the local absorbed laser intensity, and $w$ is the beam radius. Because equation 4.47 is a velocity normal to the surface of the leading edge of the keyhole, it may have components parallel and perpendicular to $v$. Thus, $v_d$ may differ from the weld speed $v$ and depends on depth within the keyhole. Under certain conditions, the component of $v_d$ in the direction of welding exceeds the weld velocity, so the keyhole surface recedes from the laser beam. This is postulated to be a source of weld instability (Matsunawa et al. 1996). $v_d$ also may be negative; this occurs when the keyhole leading surface moves toward the laser beam. Because this leads to enhanced vaporization, the ablation pressure will rise, as will $I_a$, causing the surface to again recede. This sequence of events constitutes a periodic process and may be related to the appearance of certain characteristic frequencies in acoustic and optical signals. The pressure associated with this process is calculated to lead to high melt velocities with the ejection of liquid from the front of the keyhole. This gives rise to waves of liquid that move along the leading edge of the keyhole. These waves have been observed in images of the keyhole during penetration welding (Arata 1987) and likely result in strong convective cooling effects. The flow fields in the laser-generated keyhole were simulated by Schulz et al. (1996). Additional theoretical treatments of the motion of the liquid melt produced during penetration welding with lasers can be found in the articles by Andrews and Atthey (1976), Dowden, Davis, and Kapadia (1983, 1985), Postacioglu et al. (1987), Postacioglu, Kapadia, and Dowden (1991), Schuöcker (1991), Lambrakos et al. (1991), Williams et al. (1993), Uddin and Watt (1994), Suchuöcker and Kaplan (1994), Ducharme et al. (1994), Colla, Vicanek, and Simon (1994), and Kar and Mazumder (1995). A comprehensive review was published by Mazumder (1991). The complex problem of liquid motion is closely related to the formation of the weld and ultimately controls resulting weld characteristics. However, detailed modeling is, of necessity, possible only with many simplifying assumptions and so is of limited predictive power.

## 4.7  HEAT TRANSFER IN LASER WELDING

Calculation of the temperature field in the workpiece during laser welding can provide useful information concerning the extent of the heat-affected zone and other characteristics of the weld; therefore extensive literature exists on this subject. Reviews by Gagliano and Zaleckas (1972), Ready (1971), Duley (1976), and Rykalin, Uglov, and Kokova (1978) cover early work, particularly related to heat conduction in spot welding and in conduction-limited welds. A more recent review by Mazumder (1991) and the book by Steen (1991) provide useful updates on this subject.

A fundamental advance in the modeling of heat transfer in penetration welding with lasers was published by Swift-Hook and Gick (1973) and has formed the basis for subsequent studies involving a more precise formulation of boundary conditions.

In the Swift-Hook and Gick (1973) formalism, based on a moving line heat source, the geometry of the keyhole itself is undefined other than as a cylinder of infinite thermal conductivity that delivers the incident laser radiation into the material. This simplifies the boundary conditions to the point at which a standard moving heat source solution of Carslaw and Jaeger (1959) can be used. The following simple relationship is derived between $M$, the normalized laser power per unit depth, and $Y$, the normalized width of the weld produced:

$$Y \sim 0.484M \qquad (4.48)$$

when $vw/2\kappa \gg 1$ and

$$Y \sim \exp[1.50 - 2\pi/M] \qquad (4.49)$$

when $vw/2\kappa < 1$. In these expressions $Y = vW/\kappa$ and $M = P/dKT_m$, where $W$ is the total width of the weldment and $d$ is the sheet thickness. A full penetration well is assumed.

The high-speed limit occurs in a material such as mild steel for $v$ of $>8$ m/min when one assumes a beam radius, $w$, of 150 $\mu$m. Thus, in many cases of interest, the simple expression given by equation 49 is likely a reasonable approximation. For example, with $\kappa = 0.1$ cm$^2$/sec, $W = 5 \times 10^{-2}$ cm, $K = 0.4$ W/cm °C, $T_m = 1600$ C, and $d = 0.1$ cm, $Y = 0.5$ v and $M = 0.016$ $P$. With $P = 3$ kW, the predicted welding speed from equation 4.49 is v $= 7.9$ cm/sec, or 4.7 m/min. This result assumes that all the laser power is converted to heat in the workpiece. Equation 4.49 therefore is a useful approximate relation for estimating welding speed in thin sheets and can be applied to obtain a rough estimate of welding parameters when care is taken to ensure that the low-speed limit is a valid assumption.

Unfortunately, such a simple route to the prediction of weld parameters is not usually available when more complex boundary conditions are included in the calculation of heat transfer. Indeed, the more "realistic" the model adopted in terms of including such terms as keyhole geometry, plasma absorption, and melt motion, the less flexible and the less amenable to convenient calculation are the solutions. Although modeling of heat transfer in laser welding is a useful exercise on its own, the predictive powers of most models are limited. However, the simulation software developed by Schulz et al. (1996) has been shown to have excellent predictive power in the calculation of weld parameters.

A summary of the literature on the subject of modeling of heat transfer, melt properties, and their relation to weld characteristics is given in Table 4.3. This is not a comprehensive list, but it serves to illustrate the range of boundary conditions considered and the results obtained. All models are heavily dependent in their quantitative predictions on assumed values of thermophysical parameters, some of which are poorly known over the temperature range of interest.

It is difficult to quantitatively evaluate the accuracy of these models because many of the calculated parameters can be valued only in terms of the effect they have on the final appearance and properties of the weld. This is characterized by such factors as penetration depth, weld profile, and microstructure. The shape of the keyhole both on the leading edge and on the trailing edge can sometimes be observed through radiographic imaging (Arata 1987) or in simulations, but detailed information on morphology that is predicted in various models cannot be verified in general. On the other hand, the appearance of the weld pool and its dynamics, as well as the top of the keyhole, can be readily observed and related to optical, acoustic, and plasma signals. This offers the possibility in certain cases of direct comparison between theory and experiment.

Some insight into the calculational approach to solving the complex problem of a moving keyhole of arbitrary shape in a penetration welding system can be obtained from Figures 4.21 through 4.23. Figure 4.21, which is from Kaplan (1994), shows the flow diagram leading to the conversion of laser power to heat within the keyhole. It includes the effect of plasma shielding, Fresnel absorption at the walls, multiple reflections, and plasma absorption within the keyhole. This formalism was used in determining the energy balance that gives rise to the keyhole shape shown in Figure 4.20.

An iterative approach was used by Matsuhiro, Inaba, and Ohji (1992) to calculate weld profile. The first step is an estimation of the thermal field incurred by a surface

**TABLE 4.3. Theoretical Models for Heat Transfer in Penetration Welding**

| Author | Model | Parameters Modeled |
|---|---|---|
| Swift-Hook and Gick (1973) | Line source, thin sheet | Analytical prediction of weld speed or power required |
| Andrews and Atthey (1976) | Stationary laser beam, infinite liquid | Prediction of hole depth and profile with gravity and surface tension effects included |
| Klemans (1976) | Radial heat transfer from moving cylindrical keyhole | Liquid flow around keyhole, energy balance in keyhole, keyhole shape |
| Cline and Anthony (1977) | Moving finite line source, exponential dependence on length, semi-infinite solid | Penetration depth vs. absorbed laser power |
| Mazumder and Steen (1980) | Moving Gaussian beam, plasma absorption in keyhole, effect of gas jet, finite sheet | Temperature profile vs. position |
| Chande and Mazumder (1984) | Moving workpiece, Gaussian beam, temperature-dependent thermal constants, convection losses due to gas flow | Shape of melt pool, center line cooling rate |
| Dowden, Davis, and Kapadia (1983, 1985a, 1985b) | Moving cylindrical keyhole, isothermal boundary conditions, mass flow around keyhole, finite sheet | Isotherms, shape of molten region, welding speed vs. laser power, temperature of keyhole surface |
| Dowden et al. (1987) | Finite keyhole with no absorption at walls, heat transfer via plasma, tapered cylindrical geometry, slow scan speed | Shape of open and blind keyholes |
| Postacioglu et al. (1987) | Moving tapered keyhole in sheet of finite thickness, pressure variation with depth, liquid flow | Shape of weld bead |
| Steen and Li (1988) | Keyhole modeled as combination of moving point and line sources, sheet has finite thickness | Weld cross section including bead |
| Akhter et al. (1989a) | Moving point and line source, overlapping sheets of finite thickness, power distributed with depth | Weld cross section for lap weld including galvanized sheets |
| Postacioglu, Kapadia, and Dowden (1989, 1991) | Stationary cylindrical keyhole, semi-infinite medium, Marangoni convection in weld pool with thermal conduction | Thermocapilliary flow in weld pool, oscillations of weld pool |
| Kapadia, Ducharme, and Dowden (1991) | Plasma and Fresnel absorption in tapered keyhole | Weld profile |
| Lambrakos et al. (1991) | Time-dependent energy and momentum boundary conditions at keyhole surface, numerical model | Velocity field, temperature profile |

(continued)

**TABLE 4.3.** *(continued)* Theoretical Models for Heat Transfer in Penetration Welding

| Author | Model | Parameters Modeled |
|---|---|---|
| Ducharme, Kapadia, and Dowden (1992) | Combines keyhole plasma and Fresnel absorption in consistent way, no convective terms | Analytical solution to final weld pool shape, keyhole profile |
| Kar, Rockstroh, and Mazumder (1992) | Two-dimensional axisymmetrical model for keyhole including multiple reflections and shear stress–induced liquid flow, stationary keyhole | Keyhole depth and taper, recast layer |
| Matsuhiro, Inabo, and Ohji (1992) | Uniform beam, line source of finite depth, iterative fit, noncircular keyhole cross section | Weld pool profile transverse and longitudinal section |
| Beck, Berger, and Hugel (1992) | Iterative approach subject to pressure balance in keyhole and vapor flow | Effect of laser beam parameters on welding depth and keyhole parameters |
| Ducharme, Kapadia, and Dowden (1993) | Uses Ducharme, Kapadia, and Dowden (1992) keyhole model, time-dependent laser excitation | Collapse of keyhole in relation to infrared and ultraviolet emission |
| Simon, Gratzke, and Kroas (1993) | Cylindrical keyhole with time-dependent energy flux at surface | Temperature cycle, collapse of keyhole, width of weld seam |
| Ohji et al. (1994) | Finite difference calculation, iterative calculation of keyhole geometry | Shape of weld pool, shape of weld cross section, condition for keyhole formation |
| Mueller (1994a) | Combined thermal and pressure balance, blind keyhole | Keyhole depth vs. laser power for plastics |
| Ducharme et al. (1994) | Follows Ducharme, Kapadia, and Dowden (1992) but includes convection of liquid by keyhole, thin sheets | Shape of weld pool, power dependence |
| Trappe et al. (1994) | Finite element calculation, general keyhole shape, balance of ablation pressure and surface tension | Shape and location of keyhole as functions of laser power and weld speed |
| Kaplan (1994) | Point-by-point balance of energy transfer in keyhole, moving line source, melt flow past keyhole | Weld depth vs. laser power and weld speed, keyhole shape |
| Colla, Vicanek, and Simon (1994) | Flow past cylindrical keyhole, heat flow into workpiece | Temperature distribution, shape of weld pool |
| Kar and Mazumder (1995) | Gaussian beam, finite length of keyhole, Marangoni convection, momentum and energy balance at liquid solid interfaces, scanning beam | Shape of keyhole, shape of weld pool, velocity components, isotherms in liquid |
| Avilov, Vicanek, and Simon (1996) | Heated cylinder in a stream of fluid | Thermal diffusion of hydrogen in vicinity of keyhole |
| Schultz et al. (1996) | Absorption with multiple reflections, heat conduction in liquid and solid, melt and gas flow, heat conduction in plasma, capilliary dynamics | Flow field, absorbed power, weld depth, seam width |

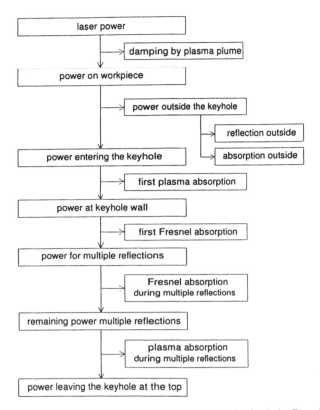

**Figure 4.21.** *Flow chart for estimation of laser power delivered to keyhole. From Kaplan (1994).*

heat source of constant intensity. The shape of the isotherms calculated at this stage is the first approximation to that of the weld pool.

In the second stage, the keyhole is allowed to form through vaporization and achieve a balance due to recoil pressure and surface tension. The maximum value of the keyhole depth is limited to the depth of the weld pool. Once the keyhole has been established, it acts as a line source of finite depth, and the temperature distribution is recalculated (stage 3). This solution then is used to recalculate the weld pool depth and the new depth of the keyhole. The overall procedure is repeated until convergence is obtained (i.e., the incremental temperature change at a given point becomes negligible). After this stage, the shape of the keyhole and weld pool have been determined both transverse to and in the direction of the laser weld.

Perhaps the most useful simulations are those that lead to a straightforward prediction of easily measured weld parameters. This is the case with the model of Swift-Hook and Gick (1973) and with some later calculations. In particular, the model of Cline and Anthony (1977) yields a set of curves that provide a good estimate of penetration depth versus laser power and welding speed.

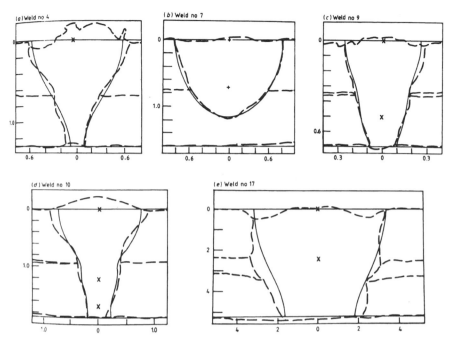

**Figure 4.22.** Lap welds between Zn-coated steel sheets. All dimensions are mm. The position of the point source is indicated by x. No line source is present in (b). Calculated efficiencies are 0.19 and 0.42 for welds (a) and (b), respectively. The scan speed was 28.6 and 32.6 mm/sec for (a) and (b), respectively. From Akhter et al. (1989).

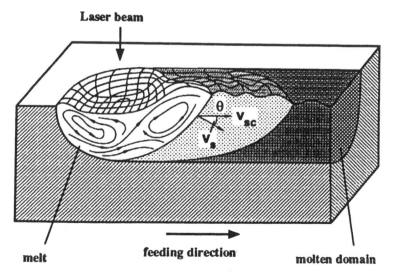

**Figure 4.23.** Schematic representation of melt pool dynamics in conduction welding. From Pirch et al. (1996).

## 4.8   WELD POOL AND WELD PROPERTIES

Details on the morphology of the weld bead and its relation to laser power, scan speed, and intensity for several metals are readily available from the plots calculated by Ohji et al. (1994) based on the calculations of Matsuhiro, Ihaba, and Ohji (1992). An excellent simulation of bead geometry in thin sheets and for lap welds between thin sheets is available from the calculations of Akhter, Watkins, and Steen (1989), based on a combined line and point source formalism. Figure 4.22 shows a comparison of calculated and observed weld cross sections in lap welds between Zn-coated steel sheets. Excellent agreement is obtained, suggesting that the relatively simple point and line source model may have useful predictive powers in welding applications of this kind.

Thermal calculations and calculation of the fluid flow fields in a weld pool are also useful in predicting microstructure. The results of such a calculation in the conduction laser weld of high-Mn stainless steel were reported by Paul and DebRoy (1988). These calculations are used to predict weld pool temperature, the spatial dependence of cooling rate within the melt pool, and velocity profiles during and after exposure to laser radiation. The surface profile after solidification and secondary dendrite arm spacings also were estimated from the model. A summary of other theoretical work on heat transfer and fluid motion in conduction welding is given in Table 4.4.

These calculations show that the initial effect of melting is to produce an expansion of the surface under the laser beam because of the density change associated with the formation of the liquid; Kreutz, Ollier, and Pirch (1992) estimate this to be ~10% of the depth of the melt. The thermal gradient causes a surface tension–driven flow of liquid away from the center of the laser beam. This flow is toward the edge of the melt pool for a negative surface tension coefficient and then down the walls to the bottom of the melt. An effect of this process is to transfer heat from the top surface of the weld pool. For a positive surface tension coefficient, the flow occurs in a surface layer from the outer part of the melt pool toward the center. This flow then is directed into the center of the melt pool and down toward its bottom. In either case, a depression is formed under the laser beam (Figure 4.23), and vortices are created within the melt pool. Microconvection cells occur near the solid/liquid interface due to the negative pressure arising on solidification.

These processes are characterized by the dimensionless parameters given in Table 4.5 (Kreutz, Ollier, and Pirch 1992). For $v = 3 \times 10^{-2}$ m/sec, $L = 3 \times 10^{-4}$ m, and $\kappa = 10^{-5}$ m$^2$/sec, Peclet number (Pe) ~ 1, which increases into the high range only when $v \geq 0.1$ m/sec. The Reynolds number (Re) $= 2.8 \times 10^3$ under the same conditions, with $v = 2.3 \times 10^{-5}$ kg/msec and $p = 7.2 \times 10^3$ kg/m$^3$, the values for Fe. Despite this large value, convection is likely laminar (Kou and Wang 1986, Mazumder 1991).

The thermal Marangoni number (Ma) depends on the temperature coefficient of the surface tension, $\sigma' = \partial\sigma/\partial T$, which is negative for metals at high temperatures. With $\sigma' = -5 \times 10^{-4}$ N/m °C as a representative value and $\Delta T = 2000C$, one obtains Ma $= -1.3 \times 10^6$ using the above values for $v$ and $\alpha$. Ma would be reduced

**TABLE 4.4. Calculation of Heat Transfer in Conduction Welding of Metals**

| Author | Model | Parameters Modeled |
|---|---|---|
| Cline and Anthony (1977) | Gaussian surface source, constant velocity, semi-infinite solid, no heat of transformation | Temperature distribution, melt penetration depth |
| Kou, Hsu, and Mehrabian (1981) | Uniform or Gaussian source, constant velocity, semi-infinite solid, finite difference method, no convection in melt | Temperature distribution, depth of melt pool |
| Sekhar, Kou, and Mehrabian (1983) | As Kou et al. (1981) applied to alloy | Isotherms, location of "mushy" zone, cooling rate |
| Chan, Mazumder, and Chen (1984) | Two-dimensional model of convective flow and heat transfer, rectangular beam, latent heat of fusion neglected, no surface distortion | Velocity field in melt, surface temperature, width/depth of weld pool, cooling rate |
| Kou and Wang (1986) | Three-dimensional model of convection and heat transfer, circular/Gaussian source | Fusion boundary, convective pattern, temperature distribution |
| Zacharia et al. (1989) | Two-dimensional heat transfer and fluid flow, Marangoni flow | Flow field, isotherms |
| Basu and Date (1990a, 1990b) | Circular surface source, scanning beam, convective heat transfer in melt | Bulk mean temperature, Marangoni number, Nusselt number |
| Mazumder (1991) | Review of previous work, two-dimensional and three-dimensional models | |
| Gratzke, Kapadia, and Dowden (1991) | Gaussian surface source, moving workpiece, high speed, thin/thick sheet, Cline and Anthony (1977) model for conduction | Weld pool shape, behavior at high speed, isotherms, penetration depth |
| Kreutz, Ollier, and Pirch (1992) | Two- and three-dimensional stationary and nonstationary model, thermocapilliary convection | Temperature distribution, flow field, melt pool geometry, surface deformation |
| Mundra and DebRoy (1993a,b) | Stationary surface source, convection in melt, heat transfer to shielding gas, evaporative heat loss | Isotherms, velocity field in melt, alloy vaporization, effect of shield gas on vaporization |
| Konstantinov, Smurov, and Flamant (1994) | Gaussian pulsed surface source, stationary solution, but weld geometry, two-dimensional model | Temperature across seam, temperature at various depths, weld pool evolution |
| Pirch et al. (1996) | Three-dimensional heat transfer, fluid flow, surface deformation, Marangoni convection, cline focus | Isotherms, flow field, surface deformation, distribution of elements |

**TABLE 4.5. Characteristic Dimensionless Parameters Related to Conduction Welding**

| Parameter | Definition | Form |
|---|---|---|
| Peclet number | $\dfrac{\text{Convective heat flux}}{\text{Diffusive heat flux}}$ | $Pe = \dfrac{vL}{\kappa}$ |
| Reynolds number | $\dfrac{\text{Convective momentum flux}}{\text{Diffusive momentum flux}}$ | $Re = \dfrac{\pi vL}{\nu}$ |
| Marangoni number | $\dfrac{\text{Rate of convection}}{\text{Rate of conduction}}$ | $Ma = \dfrac{\sigma' L \Delta T}{\nu\kappa}$ |

$L$ = scale length (m); $v$ = scan speed (m/sec); $\kappa$ = thermal diffusivity (m$^2$/sec); $\rho$ = density (kg/m$^3$); $\nu$ = viscosity (kg/m·sec); $\sigma'$ = surface tension gradient (N/m °C); $\Delta T$ = temperature difference.

considerably if L were to be identified with the depth of the melt pool but is still large, illustrating the important role of convection within the melt on heat transfer. Indeed, convective heat transfer is likely a dominant term in most laser-generated melt pools.

The sensitivity of $\sigma'$ to surface-active impurities means that weld pool behavior and morphology are strongly influenced by the presence of such elements as sulfur and oxygen in steel (Heiple and Roper 1982, Sahoo, Collur, and DebRoy 1988, Mundra and DebRoy 1993b, Pitscheneder et al. 1996). Pitscheneder et al. (1996) found that the presence of sulfur determines weld pool behavior at high Peclet numbers. Under these conditions, enhanced penetration is obtained with a higher aspect ratio. A comparison of weld cross sections for several spot welds produced in steel containing 20 and 150 ppm sulfur is shown in Figure 4.24, in which isotherms are also given. The tendency for the weld pool to form a high aspect ratio structure is clearly evident in the sample with 150 ppm sulfur, irradiated with ~5.2 kW. The estimated value of Peclet number under this condition is Pe of $>13$, whereas it is ~1 at lower incident power.

The increased penetration in the high-power–high-sulfur weld (Figure 4.24f) can be attributed in part to the fact that $\sigma'$ becomes positive for the high-sulfur steel at temperatures in excess of 1980°K (Pitscheneder et al. 1996). Simulations predict that this temperature is achieved at the center of the weld pool when P = 5.2 kW. When this occurs, Marangoni flow will occur along the surface toward the center of the weld. This produces a flow in a downward direction from the center of the weld.

Studies of vaporization during laser welding (Collur, Paul, and DebRoy 1987, Khan et al. 1988, DebRoy, Basu, and Mundra 1991, Mundra and DebRoy 1992, 1993a, 1993b) have shown how volatile components can be lost during welding as the result of selective vaporization. Loss of volatile components can lead to significant compositional changes, the formation of pores, and changes in mechanical properties. Selective vaporization occurs as convection brings alloy components to the surface of the weld pool, where, in general, they encounter a higher temperature. At high temperatures and atmospheric pressure, vaporizing species form a Knudsen layer over the metal surface (Knight 1979), which controls the net rate of vaporization and redeposits material back onto the surface. This complex interaction was modeled

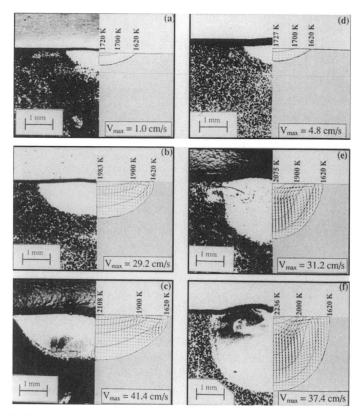

**Figure 4.24.** *Comparison of the predicted weld pool geometries with the experimental observations for the steel containing 20 ppm sulfur for laser powers of − 1900 W (a), − 3850 W (b), and − 5200 W (c) and for the steel containing 150 ppm sulfur for laser powers of − 1900 W (d), − 3850 W (e), and − 5200 W (f); irradiation time, 5 sec. From Pitscheneder et al. (1996).*

by DebRoy, Basu, and Mundra (1991) and applied to the elimination of elements from laser-welded stainless steel (Mundra and DebRoy 1993a, 1993b). A comparison of elemental vaporization rates predicted from this theory with measured rates and that estimated from the Langmuir equation is given in Figure 4.25. These vaporization rates are all much less than the equilibrium (Langmuir) rate because of the effect of redeposition and the presence of a Knudsen layer. The overall elemental vaporization rate also is found to be inhibited when a plasma is formed. Mundra and DebRoy (1993a,b) suggest that this may be due to a charging effect in which the liquid metal surface acquires a negative charge as electrons drift back from the plasma. This negative charge then acts to draw positive ions from the Knudsen layer, enhancing redeposition and reducing the net vaporization rate.

A parametric study of the effect of laser power, welding speed, and shielding gas flow on the rate of vaporization of Mn from high-Mn stainless steels has shown that Mn is preferentially lost during low-power conduction welding conditions

(Khan, DebRoy, and David 1988). Although loss of Mn was also seen in penetration welding, the relative loss under these conditions was less. The loss of Mn was slightly enhanced by a high flow of shielding gas. In all cases, the size of the weld pool was found to be an important factor in the loss of Mn, with the largest loss of Mn occurring when the melt pool was smallest (i.e., under low-power conduction limited welding conditions). This is likely due to the lack of efficient convective cooling in small melt pools.

Heating and cooling rates, as well as the rate of vaporization, can be controlled to some extent by tailoring the time dependence of the laser pulse in pulsed laser welding of metals. The effect of pulse shaping on such weld properties as penetration depth, porosity, weld area, cracking, and bead shape have been reported by Milewski, Lewis, and Wittig (1993), Katayama et al. (1993), Matsunawa et al. (1992), Matsunawa (1994), and Bransch et al. (1994).

Milewski, Lewis, and Wittig (1993) used the multiplexed output from three 400-W Nd:YAG lasers to create temporal beam intensity pulse profiles, as shown in Figure 4.26. This configuration was used to investigate the weldability of 2024-T3 aluminum, particularly the tendency for hot tearing and pore formation. It was found that the lower thermal gradient created with the high-duty cycle configuration resulted in reduced porosity and no evidence of cracking.

A comprehensive experimental study by Matsunawa (1994) using a pulse-shaped Nd:YAG laser to weld Al5083 concluded that control over the solidification rate through the use of the waveform shown in Figure 4.27 yielded optimal resistance to hot cracking. This occurred through reduction in the volume of the mushy zone at the top of the weld due to a repetitive melt-solidification cycle. With this technique, hot cracking was reduced but not eliminated. It was suggested that a longer laser pulse would remove this limitation. A study of spot welding in 3108 steel showed

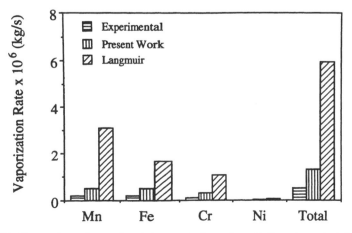

**Figure 4.25.** Comparison of experimental vaporization rates with the rates calculated from the Langmuir equation and from the model of Mundra and DebRoy (1993a).

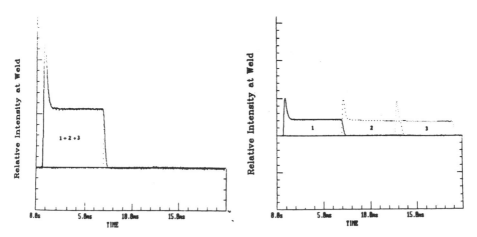

**Figure 4.26.** *Temporal intensity profiles formed by multiplexing the output of three 400-W Nd: YAG lasers. From Milewski, Lewis, and Wittig (1993).*

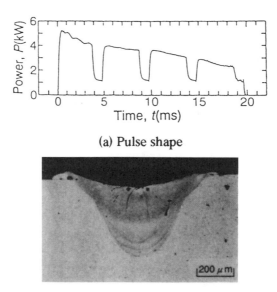

(a) Pulse shape

(b) Bead cross section

**Figure 4.27.** *Effective pulse shape to reduce hot cracking in A5083. From Matsunawa (1994).*

that the addition of a secondary, weaker laser pulse after the primary welding pulse was effective in reducing solidification cracking.

A quantitative analysis of the effect of pulse shaping on the structure, morphology, and defect content of pulsed Nd:YAG spot welds in AISI 304 stainless steel by Bransch et al. (1992) examined the connection between irradiation history and weld properties. Cross sections of spot welds were obtained, and the melt, crater, and porosity areas were measured in addition to the penetration depth. These parameters were related to the average laser intensity per pulse. Temporal pulse shapes included rectangular, rectangular with leading spike, ramp-up, and ramp-down. The overall pulse length was varied between 2 and 16 msec. A range of conduction and penetration mode welds was produced. It was found that pulse shaping had little influence on weld dimensions or weld quality, as typified by porosity, for conduction mode welds, but that when the intensity was sufficient to initiate keyhole welding conditions, the weld pool was significantly larger with ramp-up and ramp-down pulses. The porosity was largest in keyhole welds with ramp-up pulses and least with ramp-down pulses. This conclusion is similar to that reached by Matsunawa (1994) and arises because a ramp-down pulse allows the keyhole to retract from the melt pool without closing at an intermediate depth and allowing gas to be trapped on solidifiction. Conversely, when a ramp-up pulse is used, the keyhole continues to grow with time and then is abruptly terminated at the end of the laser pulse. This yields high penetration but permits gas to be trapped along the keyhole as it closes.

Studies of the relation among crater area, weld porosity, and melt area in spot welds–produced stainless steel with Nd:YAG pulses of varying durations show a complex transition between the formation of a keyhole and the creation of a weld pool. As discussed in Section 4.4, these processes occur in two distinct stages, with drilling preceding lateral heat transfer. Formation of a keyhole occupies the first ~2 msec after initiation of heating and is enhanced by the presence of a leading edge spike in the intensity profile. After ~2 msec, the keyhole has been created, and lateral heat transfer to form an extended melt begins. This liquid partially fills the keyhole so the crater area is diminished with longer pulse durations.

## 4.9  SYSTEM RESPONSE FUNCTION IN LASER WELDING

As we have seen, a wide range of different but related physical phenomena come into play in laser welding. This suggests that it should be possible to identify some general rules for laser welding that would apply generically to the laser material interaction and determine the outcome of a particular operation. Such rules should offer insight, in a semiquantitative manner, into the way in which an optimal welding regimen could be produced. In particular, they would define an operational diagram and define a range of system parameters. Some general rules for laser processing were outlined previously (Duley 1996).

The basic assumption in deriving rules for laser welding (and laser materials processing in general) is that a system exposed to laser radiation will modify its properties so as to minimize the effect of that interaction. At low incident laser

**TABLE 4.6. Rules for Laser Materials Processing**

1. All physical systems, when exposed to intense laser radiation, will transform so as to minimize their interaction with the incident radiation field.
2. The response of a physical system in adapting to an applied laser field occurs sequentially through a variety of discrete stages.
3. A system will transform so as to attempt to establish an equilibrium between excitation and deexcitation at the point of highest intensity.

intensity, this accommodation can take place through conduction of heat away from the irradiated area with little change in material properties. At higher incident laser intensity, this is accomplished through changes in morphology, such as through development of a hole in the region of laser impact. At still higher intensity, vaporization and plasma formation become important, and self-shielding may occur. These transformations occur sequentially as various accommodating mechanisms are used, with the greatest change appearing at points at which the incident intensity is highest. Some general rules that express this philosophy are given in Table 4.6.

When laser radiation is incident on a metal, the metal responds by attempting to establish a new equilibrium that is compatible with the applied laser field. At low laser intensities, this new equilibrium is established via conduction of heat away from the irradiated area. At higher incident laser intensity, equilibrium is achieved through morphological changes in the surface and through convective cooling as liquid circulates under the laser focus. In all cases, equilibrium involves a balance between the heating and cooling rates over all irradiated parts of the system. Changing the surface morphology, for example, through the development of a concavity or hole acts to lower the effective incident laser intensity and, therefore, the cooling requirements at the surface of the metal.

A system response function can be formulated in terms of a surface cooling rate. An increase in response, then, implies that the system has evolved a mechanism for higher surface cooling. This evolution will follow rules 1 and 2 as outlined below: namely that the system will adjust itself to minimize its interaction with the applied laser field and that the accommodation process will involve a hierarchy of sequential adaptation processes.

The requirement of rule 3, that the system response is driven by an attempt to establish equilibrium between excitation and deexcitation at the point of maximum intensity, suggests that the system response function $R$ can be defined as follows:

$$R = I_c(\text{max})$$

where $I_c(\text{max})$ is the maximum value of the cooling rate established at the boundary of the system. This boundary is defined to include all areas exposed to laser radiation. Because $I_c(\text{max})$ is a cooling rate, the units for $R$ are watts/per centimeter squared. It is important to note that $R$ is evaluated at the interface between the system and the surrounding medium.

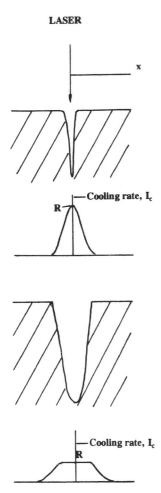

**LASER**

x

—Cooling rate, $I_c$

R

—Cooling rate, $I_c$

R

**Figure 4.28.** (Top) spatial dependence of cooling rate during drilling. (Bottom) at equilibrium.

Figure 4.28 shows the spatial dependence of the cooling rate $I_c$ during laser drilling of an opaque material. $R$ then is the value of $I_c$ at the center of the hole. As drilling proceeds to an equilibrium depth (Figure 4.28b), the cooling rate becomes more uniform over the boundary of the hole, and $R$ decreases and is constant over much of the hole profile.

$R$ is determined by the ability of the system to conduct heat away from an irradi-ated surface and, when liquid is present at that surface, to conduct heat through convective heat transfer away from interface. In addition, changes in surface mor-phology will influence $R$.

The identification of a system response function, $R$, during materials processing and, specifically, welding with laser radiation may provide a useful semiquantitative parameter in the optimization of laser materials processing techniques. As $R$ is de-fined as the cooling rate *within* the sample at the point of greatest laser heating, it indicates when the sample will change its physical, chemical, or morphological state during laser processing. Such changes must occur when $R$ reaches certain threshold values.

The simplest illustration of this occurs during laser spot heating of a solid. $R$ increases with increasing laser intensity due to the establishment of a thermal gradient within the sample. However, when $R = R_m$, the surface at which $R$ is evaluated has reached $T = T_m$ and melting occurs. For this simple example, the conditions $T \rightarrow T_m$ and $R \rightarrow R_m$ provide equivalent descriptions of the system response to laser heating. Both require a change in system behavior at $I_m$.

For $T_m < T < T_v$, temperature alone does not provide a complete description of system response because both convection and conduction become important. Of course, a comprehensive knowledge of the temperature field within the irradiated system together with a three-dimensional solution to the Navier-Stokes equations would provide the information required. However, such complete information about the system, even if it were easy to obtain, would be of limited value unless it could be incorporated into a predictive model for system response.

In estimating the response function $R$, one makes no attempt to describe in detail the behavior of the system other than that at the point at which the heating rate is greatest. $R$ does, however, encompass the overall effect of laser heating because it includes convection in the melt and the role of any changes in surface morphology. A reduction in $R$ when $I = I_v$ then reflects the fact that a hole has been opened up in the material, which enhances both convective and conductive cooling modes. Similarly, oscillations in $R$ in the plasma-shielding regimen reflect a reduction in the required cooling rate within the sample as the incident laser intensity is intermittently reduced.

To illustrate, a plot of $R$ versus log I, in which I is incident laser intensity, is shown in Figure 4.29 for spot heating of a metal surface. Conduction welding is

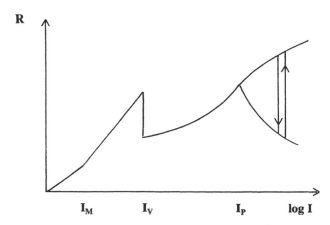

**Figure 4.29.** *System response function for laser drilling of metal with formation of melt phase. R is the response function, $I_m$ is intensity for melting, $I_v$ is vaporization threshold, and $I_p$ is the plasma threshold intensity.*

present for $I_m < I < I_v$, and the onset of keyhole welding occurs at $I = I_v$. The discontinuity in $R$ at $I = I_v$ arises as the keyhole opens, spreading the incident radiation over a larger area. Convective cooling within the melt surrounding the keyhole increases $R$ for $I_v < I < I_p$, where $I_p$ is the plasma threshold. Above $I_p$, the system oscillates between two values of $R$ corresponding to no plasma shielding (top curve) and plasma shielding (bottom curve). This behavior is characteristic of a chaotic system. A discussion of the chaotic response during laser welding was given by Otto, Geisel, and Geiger (1996).

# 5

# *Metallurgical Aspects of Laser Welding*

## 5.1 WELD THERMAL CYCLE

The temperature versus time at a particular point in a metal during and after welding is a critical parameter in determining such factors as microstructure, heat-affected zone (HAZ), and tempering. This thermal cycle can be calculated from models of the laser welding process in either conduction or keyhole welding modes (Mazumder and Steen 1980a, Metzbower et al. (1994) or measured with the use of probes (Aoh, Kuo, and Li 1992, Metzbower et al. 1993, Martukanitz, Howell, and Pratt 1992, Gilath, Signamarcheix, and Bensussan 1994). The general form of the thermal cycle is as shown in Figure 5.1, in which a rapid rise to a peak value of temperature is followed by a quasiexponential decay. For points near the laser focus or the keyhole as it scans over or through the material, the rise time of the initial increase is

$$\tau_1 \sim \ell^2/\kappa, \qquad (5.1)$$

where $\ell$ is the distance of the closest approach of the heat source to the point at which temperature is recorded. For a point located on the centerline of the heat source as it traverses, $\tau_1$ becomes $\tau_1 \sim w^2/\kappa$ where $w$ is the radius of the source. With $\kappa \sim 0.1$ cm$^2$/sec, $\ell = 1$ mm, and $w = 0.2$ mm, values for $\tau_1$ are 0.1 and 0.04 second, respectively.

An analytical expression for $T(t)$ for a moving surface heat source can be obtained from the result derived by Ashby and Easterling (1984):

$$T(r,\ t) - T_0 = \frac{P}{4\pi \text{v}Kt} \exp\left(-\frac{\text{v}^2}{4\kappa t}\right) \qquad (5.2)$$

for a thick slab and

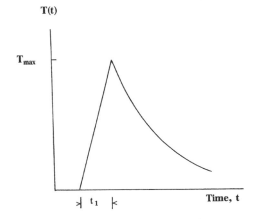

**Figure 5.1.** Schematic for the thermal profile in laser welding.

$$T(r, t) - T_0 = \frac{P}{vd(4\pi K_p ct)^{1/2}} \exp\left(-\frac{r^2}{4\kappa t}\right) \tag{5.3}$$

for a sheet of thickness $d$. The laser power is P, the scan speed is v, and $T_0$ is the ambient or preheat temperature. For a thick sheet, the maximum temperature $T_{max}$ is reached at a time

$$\tau_1 = \frac{r^2}{4\kappa}, \tag{5.4}$$

and the rate of change of temperature, dT/dt, is

$$\frac{dT}{dt} = \left(\frac{P}{4\pi v K t^2} \exp\left[\frac{-r^2}{4\kappa t}\right]\right)\left(\frac{r^2}{4\kappa t} - 1\right). \tag{5.5}$$

Equations 5.2–5.5 assume an instantaneous heat source and therefore are approximate only in their prediction of the time dependence of T and $dT/dt$. However, given the other limitations of this model, reasonable agreement is obtained with experimental data (Easterling 1983).

Using numerical values for Fe, with $P = 3kW$ and $v = 5 \times 10^{-2}$ m/sec:

$$\frac{dT}{dt} = \frac{117}{t^{1/2}}\left[\frac{6.25 \times 10^{-3}}{t} - 1\right]\exp\left[\frac{-6.25 \times 10^{-3}}{t}\right], \tag{5.6}$$

so $T_{max}$ is reached at a time $\tau_1 = 6.25$ m/sec.

Thermal cycles for locations within the workpiece valid under conditions of penetration welding require a numerical solution of the heat equation. An example of the results of such a calculation (Mazumder and Steen 1980a) yields a similar func-

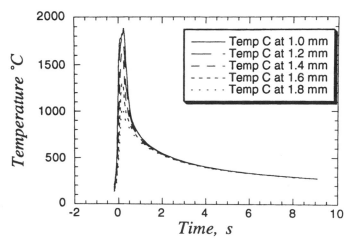

**Figure 5.2.** Calculated thermal profiles at various locations across the fusion zone during weld-ing of HSLA80 steel. The weld speed was 11 mm/sec at P = 12 kW. From Metzbower et al. (1994).

tional form to that of equations 5.2 and 5.3. Additional examples are given by Metzbower (1994) (Figure 5.2).

Measurements of $T(t)$ carried out using thermocouples (Aoh, Kuo, and Li 1992, Gilath, Signamarcheix, and Bensussan 1992, Martukanitz, Howell, and Pratt 1992) are similar in profile to those shown in Figure 5.2, although $T_{max}$ and other param-eters depend on welding speed, laser power, material constants, and so on.

The thermal cycle may be used to infer the effect of welding on the region of the HAZ adjacent to the weld. Because heating rate in this region determines grain growth through the dissolution and coarsening of precipitates, it determines the thermal history of the HAZ before the onset of cooling. The final microstructure is determined by this prior treatment and the cooling rate (Lancaster 1993).

A finite element technique was developed by Yang, Hsu, and Albright (1993) to calculate the thermal cycle at various points in butt welds between steel sheets of different gauge. Welds were produced under typical tailor blanking conditions (e.g., $P \sim 3$ kW and v up to 0.2 m/sec). Cooling rates estimated from these thermal cycles were found to be in the range of 700–3500°C/sec, with the highest values obtained near the edge of the fusion zone in a butt weld between two sheets of the same gauge. Simulation of the boundary of the HAZ and fusion zone yielded good agreement with experiment.

## 5.2 THE HAZ

### 5.2.1 Steels

The HAZ is adjacent to the weld and displays the effect of temperature cycling to a peak temperature, $T_{max}$, which is less than the melting point but may be sufficient to initiate other transformations. The cooling rate in the HAZ may approach 1000°C/sec but varies with location, as does $T_{max}$. For a linear butt weld in thin plate, the width of the HAZ is comparable to that of the weld itself and is a region of altered hardness and variable microstructure.

For a thin plate, and with a simplified heat source model, the lateral dimension of the HAZ can be defined in relation to a specific transformation temperature $T'$ (Ion, Salminen, and Sun 1996):

$$r' - r_m = \frac{AP}{\upsilon d}\left[\frac{0.34}{\rho c}\left(\frac{1}{T' - T_0} - \frac{1}{(T_m - T_0)}\right)\right],\tag{5.7}$$

where $r'$ is the lateral distance from the heat source to the point $T = T'$, $r_m$ is the corresponding distance to the melt interface $A$ is absorptivity, and $T_0$ is the ambient temperature. With $AP = 3$ kW, v $= 5 \times 10^{-2}$ m/sec, $\rho c = 5.3 \times 10^6$ $J/m^3°C$, and d $= 1$ mm, one obtains

$$r' - r_m = 3.8\left[\frac{1}{(T' - T_0)} - \frac{1}{(T_m - T_0)}\right].\tag{5.8}$$

With $T'$ typically in the 800°C range for steels and with $T_m = 1600°C$, one obtains, taking $T_0 = 20°C$, $r' - r_m \sim 2.5$ mm. Values of the width of the HAZ for several steels welded with a variety of $CO_2$ laser powers, together with other parameters associated with the HAZ and the laser welding process in general, are given in Table 5.1 (Ion, Salminen, and Sun 1996). The measured width of the HAZ, $r' - r_m$ can be used in equation 5.8 to obtain an estimate of the absorptivity, $A$. Values of $A$ estimated in this way are seen to range between 0.4 and 0.9 (Table 5.1).

The hardness in the HAZ zone is dependent on the absorbed energy, $q = AP/vd$ $(J/m^2)$ and tends to decrease as q increases. This effect can be seen in Figure 5.3 for a carbon Mn steel and is due to an increase in cooling time at large values of $q$. For example, when $q = 10$ $J/mm^2$, the cooling time $\Delta t$ is 0.1 second and HV is 400. An increase in $q$ to 50 $J/mm^2$ results in $\Delta t$ of 3.5 second and HV of 230. This decrease in HV is due to the increased volume of bainite and a reduction in the volume of martensite. Fits to the experimental data using the theoretical models of Ion, Salminen, and Sun (1996), Ion, Easterling, and Ashby (1984), Terasaki (1981), and Yurioka, Okumura, and Kasuva (1987) also are shown in Figure 5.3 and are seen to yield good agreement with these data. The poorest agreement between predicted hardness and observational data was obtained in thermomechanically processed steels, but even in this case, the discrepancy was limited to ~20%. The theoretical models can then be used with some reliability in predicting the maximum hardness in the HAZ of steels of various compositions.

Because the thermal cycle depends on the distance away from the fusion zone interface, composition, microstructure, and mechanical properties vary over the HAZ (Metzbower 1990). An example of the hardness variation across the HAZ in the $CO_2$ laser–welded high-strength steel HSLA80 is given in Figure 5.4 and shows

**TABLE 5.1. Properties of HAZ in Several Steels after $CO_2$ Laser Welding**

| Welding Parameter | | | | HAZ Width (mm) | | | | HAZ Hardness (HV1) | | | | Absorptivity (%) | | | |
| Plate Thickness (mm) | Power (kW) | Speed (m/min) | Applied Energy (J/mm²) | Fe37B | Fe52D | HSD | HSE | Fe37B | Fe52D | HSD | HSE | Fe37B | Fe52D | HSD | HSE |
|---|---|---|---|---|---|---|---|---|---|---|---|---|---|---|---|
| 4.0 | 2.5 | 0.6 | 63 | 1.36 | 1.60 | 1.45 | 1.75 | 234 | 248 | 363 | 339 | 60 | 72 | 62 | 76 |
| | | 1.4 | 27 | 0.67 | 0.85 | 0.71 | 0.77 | 340 | 348 | 413 | 407 | 69 | 89 | 71 | 78 |
| | 3.8 | 2.1 | 27 | 0.55 | 0.80 | 0.55 | 0.64 | 328 | 368 | 378 | 413 | 57 | 84 | 55 | 65 |
| | | 2.4 | 23 | 0.56 | 0.64 | 0.49 | 0.60 | 283 | 413 | 395 | 413 | 67 | 76 | 56 | 70 |
| 3.8 | 3.8 | 0.4 | 94 | 2.05 | 2.60 | 2.00 | 2.75 | 210 | 210 | 283 | 293 | 61 | 78 | 57 | 80 |
| | | 0.6 | 63 | 1.96 | 1.75 | 1.40 | 1.60 | 222 | 245 | 339 | 334 | 87 | 78 | 60 | 70 |
| | | 0.8 | 47 | 1.54 | 1.20 | 0.90 | 1.30 | 214 | 269 | 373 | 378 | 91 | 72 | 52 | 76 |
| 6.0 | 5.0 | 1.1 | 47 | 1.12 | 0.95 | 0.90 | 1.15 | 212 | 286 | 363 | 358 | 66 | 57 | 52 | 67 |
| | | 1.4 | 36 | 1.00 | 0.80 | 0.90 | 1.07 | 242 | 325 | 368 | 413 | 78 | 63 | 68 | 82 |
| 8.0 | 5.0 | 0.5 | 75 | 2.24 | 1.95 | 1.70 | 2.05 | 218 | 242 | 321 | 293 | 83 | 73 | 61 | 74 |
| | | 0.6 | 63 | 1.75 | 1.73 | 0.95 | 1.55 | 215 | 237 | 325 | 321 | 78 | 77 | 41 | 68 |

From Ion et al. (1996).

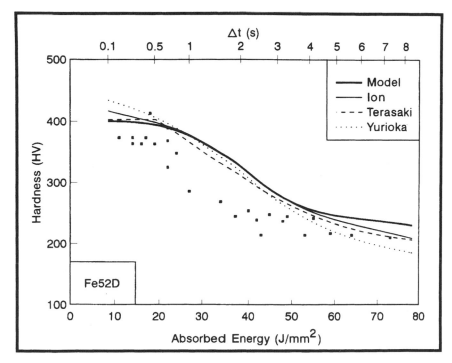

**Figure 5.3.** *Maximum HAZ hardness plotted against absorbed energy for laser beam welding of the steel Fe52D. Experimental data and the predictions of both the present model and more sophisticated models are shown. From Ion, Salminen, and Sun (1996).*

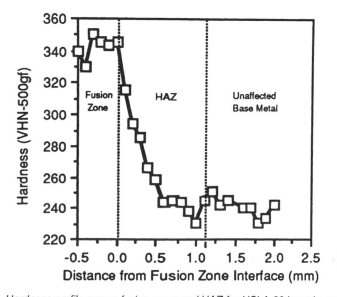

**Figure 5.4.** *Hardness profile across fusion zone and HAZ for HSLA-80 laser beam weld. From Martukanitz, Howell, and Pratt (1992). (Reprinted with permission from ASM International)*

**TABLE 5.2. Estimation of Phase Proportions at Various Distances from the Fusion Zone Interface for HAZ of Laser Beam–Welded HSLA-80**

| Distance (mm) | Grain Size* ($\mu$m) | LM | PF | AF | M(A) |
|---|---|---|---|---|---|
| 0.1 | 17 | 0.95 | — | — | 0.05 |
| 0.3 | 4 | 0.30 | 0.21 | 0.44 | 0.05 |
| 0.5 | 3 | — | 0.65 | 0.30 | 0.05 |
| 0.7 | 5 | — | 0.73 | 0.22 | 0.05 |

* Grain size at the 0.1-mm position represents previous austenite grain size.
From Martukanitz, Howell, and Pratt (1992).

the smooth transition from the high-hardness level produced within the fusion zone to that of the base metal (Martukanitz, Howell, and Pratt 1992). This transition is well correlated to variations in microstructure across the HAZ. A summary of the relative proportion to various phases observed in the HAZ under these conditions is given in Table 5.2. The proportion of lathe martensite (LM) is as expected—largest in the region closest to the fusion zone where the cooling rate is greatest. Acicular ferrite (AF) and polygonal ferrite (PF) commensurate with a lower cooling rate are dominant phases at distances of $>0.5$ mm from the fusion zone interface. Martukanitz, Howell, and Pratt (1992) estimated that the maximum value of $dT/dt \sim 300°C/\text{sec}$ at $r' - r_m = 0.5$ mm. A small amount of high-carbon austentite and retained austentite, M(A), is inferred at all positions. The grain size is largest (17 $\mu$m) close to the fusion zone and is attributable to austenite grain growth. A similar conclusion was reached by Metzbower, et al. (1994).

In a study of $CO_2$ laser welding of ASTM A-36 steel, Metzbower and Moon (1981) found that the microstructure in the HAZ was typical of refined ferrite/pearlite and that little increase in hardness was obtained except for points very close to the fusion zone. A bainite microstructure was found in the fusion zone itself. Welding conditions corresponded to energy inputs $q = 50$–$150$ J/mm$^2$. Both an absence of hardening in the HAZ and the observed microstructure are consistent with these values of q because they indicate a slow cooling rate.

A series of $CO_2$ laser welding experiments on a range of carbon steels by Hall and Wallach (1989) support this conclusion and show the key role of energy input and subsequent cooling rate on the distribution of hardness within the weld and across the HAZ. An example of the spatial variation of hardness under low-energy input (B1, 50 J/mm$^2$) and high-energy input (B5, 150 J/mm$^2$) conditions is shown in Figure 5.5. These data show that although peak hardness is obtained with low-energy input, this occurs at the center of the weld. At high $q$, peak values of hardness are limited to the edge of the HAZ near the fusion zone and the hardness within the weld is reduced.

The variation in hardness in the fusion zone and the HAZ with weld speed using high-speed $CO_2$ laser welding of two automotive steels was reported by Baysore et al. (1995) (Figure 5.6). Type A steel had a higher S, Ni, and Cr content than steel B and was less formable. It exhibited marginally higher hardness within the HAZ than that of steel B. Although the maximum hardness within the fusion zone was

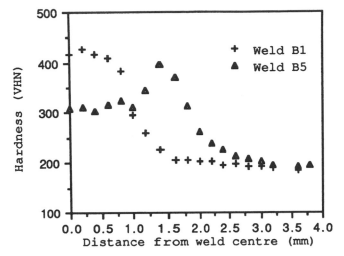

**Figure 5.5.** Hardness profile for welds B1 and B5 (low- and high-energy input, respectively) for bead-on-plate weld with 10-kW $CO_2$ laser and carbon steel. From Hall and Wallach (1989). (Reprinted with permission ASM International)

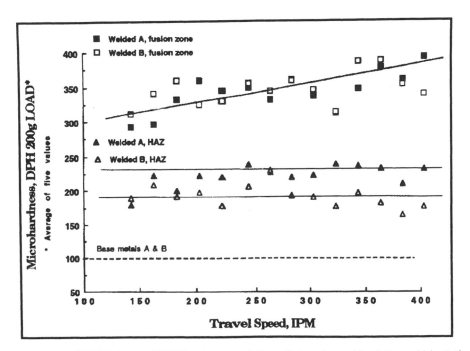

**Figure 5.6.** Weld fusion and HAZ hardness variation with travel speed for 0.8-mm-thick steel and 3–5-kW $CO_2$ laser. From Baysore et al. (1995).

found to increase linearly with welding speed, that in the HAZ was virtually independent of speed. Simulations of heat transfer in thin sheets under these high-speed welding conditions by Yang, Hsu, and Albright (1993) predict that the cooling rate can be very large (~3–5 × $10^{3}$°C/sec) at points near the fusion zone interface.

Microstructural investigations of thin stainless steel sheets welded with low-power $CO_2$ laser radiation under conditions at which $q \sim 100$ J/mm$^2$ in a conduction welding mode show that the width of the HAZ decreases slowly with welding speed for $2 \leq v \leq 8$ mm/sec (Khan 1993). At constant v, the width of the HAZ increases slightly as laser power increases. The variation in hardness across the HAZ and changes in microstructure are shown in Figure 5.7. Additional studies of low-power $CO_2$ laser butt welding of thin stainless steel sheets were reported by Vilar and Miranda (1988), Zuyao, Chu, and Bingyou (1987), and Arai et al. (1987). Arai et al. (1987) found that the microhardness was lowest at the center of the weld. Residual stresses also were highest at this point, leading to a tendency for cracking. Maximum values of $dT/dt$ were found to be in excess of $10^{3}$°C/sec at the edge of the weld zone.

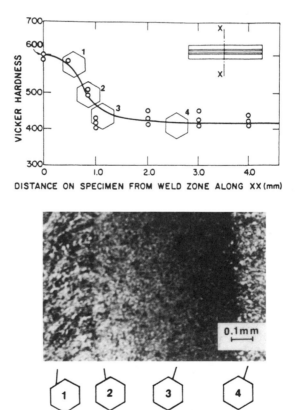

**Figure 5.7.** *Microhardness and microstructural variations in laser-welded AISI 4130 steel for laser power of 300 W and welding speed of 3 mm/sec. From Khan (1993).*

Coyle (1983) reported on a comparative study of low-power continuous-wave (CW) and pulsed $CO_2$ and pulsed Nd:YAG laser welding of thin sheets of 316L stainless steel. It was found that pulsed welding conditions yield faster cooling conditions and favor primary austentitic solidification, whereas CW welding gives solidification in an austentitic-ferritic mode.

Some additional studies of the HAZ in laser-welded steels were reported by Kujenpaa, Helin, and Moisio (1988), Kujenpaa and Moisio (1989), Liang, Zhang, and Feng (1992), and Grezev, Grigoryants, and Fedorov (1984). The effect on the HAZ hardness profile of the addition of filler wire in laser welding of H480 and H4100 (Phillips and Metzbower 1992) and in API-X80 pipeline steel (Harvey and Wallach 1992) also was evaluated. The effect of a hot wire feed in H480 is to broaden the HAZ without a change in hardness.

## 5.2.2 Aluminum Alloys

Laser welding of aluminum alloys often is accompanied by a loss of volatile elements from the weld region. This is particularly true of alloys of the 5000, 6000, and 7000 series, with loss of Mg an important factor for the 5000 and 6000 alloys and of Zn for the 7000 alloys. Non–heat-treatable alloys such as 5454 have mechanical properties that are improved by cold working and are susceptible to the heat introduced by laser welding. In particular, alloys in the 5000 and 6000 series are subject to solidification cracking and to reduced strength in the HAZ.

Heat-treatable alloys such as 7075 also are susceptible to hot cracking and solidification cracking after welding, but mechanical properties may be regenerated through postweld heat treatment. An important effect is the annealing that occurs in a narrow region in the HAZ adjacent to the weld. When -T4 and -T6 tempers are welded, the heat of welding acts to destroy the effect of heat treatment in this narrow region. This occurs because of the reversion or dissolution of $Mg_2Si$ precipitates that takes place at temperatures between ~250 and ~500°C (Grong 1994). The loss in strength may be recovered through aging at temperatures between 150 and 180°C.

The effect of laser welding on aluminum alloy properties is dependent, as in steel, on the thermal cycle. The thermal cycle in the weld pool determines the loss through vaporization of elements such as Mg, whereas that in the HAZ results in a reduction in hardness due to dissolution of precipitates. The tendency for vaporization to occur is enhanced in Al alloys because a high laser intensity is required to initiate the welding condition since these alloys have high reflectivity together with large thermal conductivity. The effect of reflectivity can be reduced to some extent by using a shorter laser wavelength (e.g., Nd:YAG).

Hardness profiles and microstructures within the weld zone and HAZ were obtained as part of some of the first studies of $CO_2$ laser welding of Al alloys by Moon and Metzbower (1979, 1983). The alloy studied was Al-5456 with a composition by wt.% of 5.25% Mg, 0.80% Mn, and 0.10% Cr. The parent material had distributed $Mg_2Al$ precipitates in an Al matrix together with insoluble particles of $Mg_2Si$ and (Fe, Mn) $Al_6$, whereas the fusion zone consisted solely of α-Al and a $Mg_2Al_3$ precipitate. In the HAZ, elongated grains and a large number of insoluble $Mg_2Si$

particles were found, together with a second-phase $Mg_2Al_3$. The resulting hardness profile showed a large reduction within the fusion zone together with an additional reduction at the fusion zone HAZ boundary. The hardness recovered to that of the parent material through the HAZ. This was accompanied by a loss of Mg from the fusion zone, but no change in Mg concentration was detected in the HAZ. The concentration of other elements, such as Mn and Cr, also were essentially unchanged in the HAZ.

The reduction in hardness in the HAZ, especially near the fusion zone boundary, must then arise through reversion of precipitates rather than through any loss of Mg. The tensile properties of these laser welds were reduced over those of the base metal. This was attributed to three factors:

1. depletion of Mg
2. loss of strain-hardened structure
3. porosity

A study of the effects of Nd:YAG laser welding on the heat-treatable alloy 2024 in the T3 temper by Milewski, Lewis, and Wittig (1993) has shown that under high pulse energy conditions, loss of strengthening precipitates in the HAZ leads to ductile failure. This can be seen in Transmission Electron Microscope (TEM) images of the base material, HAZ, and fusion zone (Figure 5.8).

Recovery of hardness in a AlMgSi alloy within the HAZ through a full age-hardening heat treatment was demonstrated by Behler et al. (1988b,c). This suggests that welding of such alloys in the soft annealed state may be appropriate.

Additional reports on microstructure and hardness variation within the HAZ for laser-welded Al alloys are given by Avramchenko and Molchan (1983) (AA 5085), Grigoryants et al. (1983) (AA 5085), Cieslak and Fuerschbach (1988) (6061, 5456, 5086), Gnanamutha and Moores (1987) (7475), Thorstensen (1989) (AA 6082), and Martukanitz et al. (1994) (5754).

Starzer et al. (1993) examined the effect of the addition of filler wire or continuous powder feed on the hardness distribution in the fusion zone and HAZ in the age-hardened alloys 6060T6 and 6082T6. A severe reduction in hardness was observed both within the HAZ and the weld zone without filler material, but a hardness reduction in the fusion zone could be reduced through the addition of AA 4043 or AA 4047 as filler. Tensile failure was found to move from the weld zone to the HAZ with increasing addition of filler material.

### 5.2.3 Titanium Alloys

Microstructure in the HAZ in Ti-6Al-4V after $CO_2$ laser welding was discussed by Mazumder and Steen (1979) and Mazumder (1983), and a comparison between microhardness variations in laser-welded Ti and Ti-6Al-4V was reported by Denney and Metzbower (1989). Microstructure and composition across $CO_2$ laser welds between Ti-6Al-4V and the metastable β-titanium alloy Beta-C were obtained by

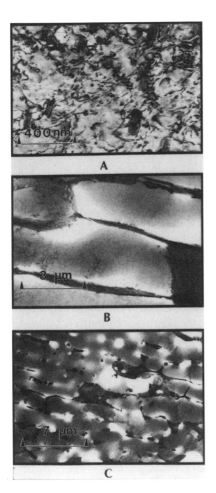

**Figure 5.8.** Bright field (BF) TEM images from HDC weld. (A) base material. (B) HAZ. (C) Fusion zone. From Milewski, Lewis, and Wittig (1993).

Liu, Baeslack, and Hurley (1994) and related to mechanical properties. Postweld aging was found to result in a significant increase in hardness within the HAZ.

Nd:YAG laser welding in a Ti-Al-Nb titanium aluminide (Baeslack, Cieslak, and Headley 1989) followed by postweld heat treatment has been shown to result in an improvement in weld ductility. The dependence of microstructure on postweld heat treatment is discussed here.

## 5.2.4 Other Alloys

The Ni-Al-bronzes for use in marine applications are subject to selective phase corrosion in and near the HAZ after arc welding. It has been suggested that this effect may be minimized through laser welding because this produces a narrower HAZ (Petrolonis 1993, Bell, Petrolonis, and Howell 1992). The dependence of the

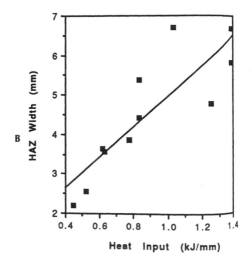

**Figure 5.9.** Width of HAZ in Cu-9Al-4Fe-4Ni, Ni-Al-bronze versus heat input P/v. Data obtained at 50% of penetration depth for 14-kW $CO_2$ laser. From Petrolonis (1993).

width of the HAZ in this material on heat input, P/v, is shown in Figure 5.9. The microhardness distribution is characterized by sharp spikes at the fusion/HAZ boundary corresponding to martensitic material. The HAZ consists of Widmanstatten $\alpha$ and martensite as the result of rapid cooling of the $\beta$ phase (Marsico et al. 1993).

### 5.2.5   Dissimilar Metals

Laser welding of dissimilar metals has been investigated since the first studies on laser welding (Charschan 1972). The microstructure for several dissimilar welds and properties within the HAZ can be found in reports by Garashchuk et al. (1969) (Ta/Mo, Ni/W), Garashchuk and Molchan (1969) (Ni/Cu, Ni/Ti, Cu/Ti), Baranov et al. (1968) (brass/Cu, mild steel/Cu, Cu/SS), Grezev et al. (1984) (steel/steel), Kujanpaa, Helin, and Moisio (1988), and Kujanpaa and Moisio (1989) (SS304/1018 steel).

### 5.3   FUSION ZONE

### 5.3.1   General

Because the fusion zone contains the region at which laser radiation is absorbed, T(t) profiles in general extend to higher temperatures than those existing in the HAZ. If a keyhole is formed or the surface temperature in a conduction weld reaches the vaporization temperature for an alloy constituent, then selective vaporization can result in the loss of minor elements. This can have a profound influence on the hardness and mechanical properties of the weldment. The volatilization of alloy constituents will occur in addition to changes in phase and the redistribution of precipitates.

The liberation of volatile elements including hydrogen can result in porosity that may extend over the size range from $\ll 1$ μm to macroscopic bubbles. This is particularly important during laser welding of Al alloys, when the evolution of hydrogen and the limited solubility of hydrogen in the melt often produce a porous weld.

Rapid cooling rates, segregation in the melt, and the presence of impurities can lead to solidification cracking in susceptible alloys. Transverse and longitudinal stresses play a major role in cracking but may be minimized through control over the time dependence of laser power input to the workpiece. Liquation cracking or hot tearing also is problematical in many materials and is initiated by grain boundary segregation and enhanced at high cooling rates.

Although the morphology and macrostructure of laser welds depend to a large extent on processes occurring during welding, including the form of the keyhole, Marangoni convection, recoil-induced liquid flow, and others, the microstructure of these welds depends on the solidification process.

A simplified representation of the weld pool shape parallel and transverse to the welding direction is shown in Figure 5.10 to illustrate the basic features of the solidification process. The solidification rate, $v_s$, is related to the welding speed, v, as follows

$$v_s = v \cos \theta, \tag{5.9}$$

where $\theta$ is the angle shown. The local solidification rate varies from the top of the weld pool to the bottom (Figure 5.10, bottom), whereas the solidification front propagates with time from the substrate toward the top of the weld pool (Figure 5.10, middle). Because solidification occurs from the solid/liquid interface into the weld pool, epitaxial growth may take place with grain growth initiated by the grains in the region of the HAZ closest to the fusion zone boundary. Individual grains will have a substructure resulting from microsegregation and are influenced by the solute content of the melt. A characteristic solidification parameter $G/\sqrt{R}$ can be defined where G is the thermal gradient in the direction of solidification and R is the rate of advance of the solidification front. The microstructure becomes more dendritic as $G/\sqrt{R}$ decreases while the spacing between dendrites increases with freezing time. This relation forms the basis for an estimation of melt cooling rates from dendrite spacing (Gilath, Signamarcheix, and Bensussan 1994). A striking example of dendritic growth epitaxial to the substrate and in the direction of heat flow can be seen in Figure 5.11. This organized microstructure gives way to a misoriented grain structure in regions close to the center of the weld pool.

The growth of specific grains will be enhanced when the heat flow direction corresponds to the appropriate crystallographic orientation. In polycrystalline materials, grains with the optimum crystallographic orientation relative to the direction of heat flow will be able to grow efficiently. The growth of dendrite tips under rapid solidification conditions was discussed in detail by Kurz, Giovanola, and Trivedi (1986), Bobadilla, Lacaze, and Lesoutt (1988), and Gilgen and Kurz (1996).

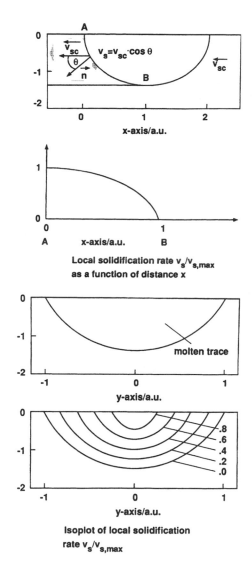

**Figure 5.10.** Top schematic represent of weld cross section in the direction of welding. Middle transverse to weld. Lower relative solidification rate. Adapted from Kreutz, Ollier, and Pirch (1992).

A study of electron beam and $CO_2$ laser welding of single-crystal Fe-15Ni-15Cr alloy (Rappaz et al. 1989) reveals how dendritic growth is related to both crystallographic orientation and thermal gradient within the weld pool. Figure 5.12a shows a transverse section through a weld produced with an electron beam at $v = 3$ mm/sec. Various regions of grain growth are identified and related to crystallographic directions within the single crystal in Figure 5.12b. Welding was carried out along the [100] direction. The top of the weld at the centerline and its root are characterized by dendritic grain growth along the [100] scan direction. At intermediate depths, the growth is preferentially along the [001] direction, whereas grain growth from

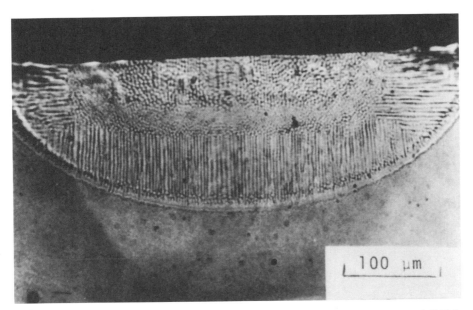

**Figure 5.11.** *Transverse section of laser-melted region in Udimet 700. From Copley et al. (1981).*

the side of the keyhole in regions below the nail head corresponds to dendritic growth along the [010] and [0$\bar{1}$0] directions.

The clear influence of weld pool geometry in determining columnar grain growth is apparent in Figure 5.12. This suggests that weld speed and its relation to thermal properties of the base material are important factors in determining grain structure in laser welds. Grong (1994) defines the dimensionless number as

$$n_z = \frac{Pv}{4\pi\kappa^2 \Delta H_m}, \tag{5.10}$$

where $\Delta H_m$ is the heat content per unit volume at the melting temperature. With this designation, large values of $n_z$ correspond to elongated or tear-shaped welds. This condition will promote straight and broad columnar grain structure with a tendency for centerline cracking. When $n_z$ is small, the weld pool tends to be elliptical or quasispherical, and grain growth is greatest at the centerline, leading to tapered and curved columnar grain structures that follow the thermal gradient from the fusion zone boundary to the centerline of the weld.

Gilath, Signamarcheix, and Bensussan (1994) reported that the primary dendrite spacing, $\lambda_1$, in laser-welded 304L stainless steel is given by the following simple relation:

$$\lambda_1 \ (\mu m) = \frac{40}{v \ (mm/sec)^{1/2}}, \tag{5.11}$$

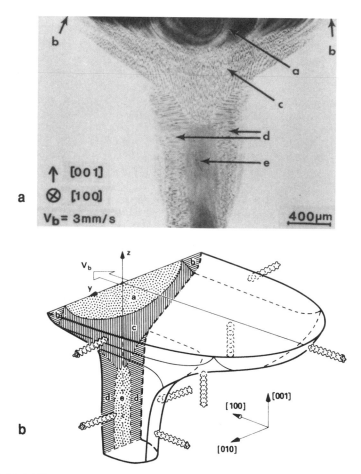

**Figure 5.12.** *(a) Transverse section photomicrograph of a Fe-15Ni-15Cr single-crystal electron beam weld made at 3 mm/sec along a [100] direction on a [001] surface. (b) Three-dimensional schematic of the weld pool and dendrite growth orientations. From Rappaz et al. (1989). (Reprinted with permission ASM International)*

where $v$ is the welding speed. The cooling rates inferred from this observation depended on laser power and scan speed but typically ranged between $10^3$ and $3 \times 10^3$ °K/sec with temperature gradients of $1-9 \times 10^2$ °K/cm.

## 5.3.2  Steels

At high cooling rates, nonequilibrium solidification processes may become important. This was discussed in a comprehensive report by Vitek and David (1994) on laser welding of austentitic stainless steels. They found that the

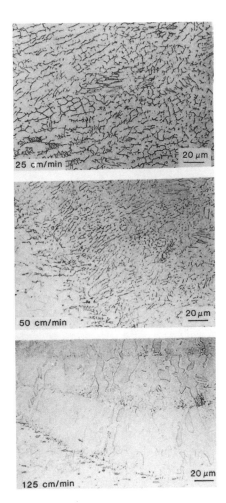

**Figure 5.13.** Microstructures in laser-welded type 304 stainless steel at three different welding speeds showing the change in solidification mode from primary ferrite at low speed to primary austenite at high speed. From Vitek and David (1994).

solidification structure strongly depends on welding speed (Figure 5.13). At v = 25 cm/min, the weld zone in 304 stainless steel shows microstructure consistent with a primary ferritic solidification structure. At 125 cm/min, the microstructure becomes essentially 100% austentitic and has an increased tendency to hot cracking. David and Vitek (1981) also found evidence for a primary austentite solidification mode in laser-welded 308 stainless steel. However, this was observed primarily at the root of the weld, where the cooling rate presumably was greatest. Further reports on the microstructure of laser welds in stainless steels were made by Coyle (1983), Vilar and Miranda (1988), Kujanpaa and David (1986), Woolin (1994), Khan (1993), Arai et al. (1987), and Zuyao, Chu, and Bingyon (1987). A detailed analysis of the relative volume fraction of ferrite and martensite microstructures as a function of position across the fusion zone in the high-strength steel HSLA80 was provided by Metzbower et al. (1994). The proportion of martensite was highest at the centerline of the weld and the boundary of the fusion zone. Thermal gradients would be greatest in these regions.

The depletion of volatile alloying elements in the weld pool of some laser-welded steels was determined by Paul and Khan (1994) in both conduction and keyhole welding modes. The loss of Mn from the weld pool was found to be significant in both modes and increased with laser power and welding speed. Modeling of these effects (Mundra and DebRoy 1993a) shows that the type and flow rate of shielding gas are important variables in the overall vaporization rate.

Weld microstructure and toughness in steels can be influenced through postweld tempering (Aoh et al. 1992, Abdullah et al. 1995). This is accomplished through passage of the defocussed laser beam along the weld. Tempering of laser-welded SAE 4130 steel with this technique resulted in the conversion of the coarse-grained martensitic structure into a finer martensite structure with small amounts of bainite. Several tempering passes were required to remove the dendritic structure in the welded material and to temper throughout the full depth of the weld.

The addition of filler material during $CO_2$ laser welding of pipeline steel to promote toughness through modification of weld metal microstructures was discussed by Harvey and Wallach (1992). The use of hot filler wire was reported by Phillips and Metzbower (1992).

### 5.3.3  Aluminum Alloys

Aluminum alloys are particularly sensitive to the loss of additive elements, and this can have a profound effect on the strength of welds, particularly those of the 5000–7000 series. A plot of melting and vaporization temperature for various alloying elements from Sakamoto, Shibata, and Dausinger (1992) is shown in Figure 5.14; Mn, Mg, Cr, and Zn are susceptible to depletion during welding.

A representative selection of weld cross sections for bead-on-plate welds in Al alloys using $CO_2$ laser radiation is shown in Figure 5.15. These blind welds have a similar shape to those produced in steels. Welds in 6061 and 7075 were subject to cracking at high welding speeds; the cracks occurred along the boundaries of dendritic crystals.

$CO_2$ laser welds in Al alloys often are characterized by a microstructure that includes coarse-grain equiaxed grains in the middle of the weld with columnar grains along the interface with the HAZ (Avramchenko and Molchan 1983, Grigoryants et al. 1983, Gnanamuthu and Moores 1987, Mazumder 1983).

The porosity in laser welds depends on a variety of factors, including the type and flow rate of the shield gas, welding speed, and focussing conditions. Surface preparation before welding is not as important (Figure 5.16). The role of hydrogen in the formation of these pores is unclear because the correct surface preparation should reduce hydrogen but data show little affect of different cleaning techniques. Rapp et al. (1993) noted that pore-free welds can be produced at

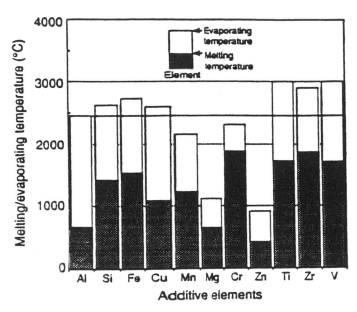

**Figure 5.14.** Melting and vaporization temperatures of additive elements in Al alloys. From Sakamoto, Shibata, and Dausinger (1992).

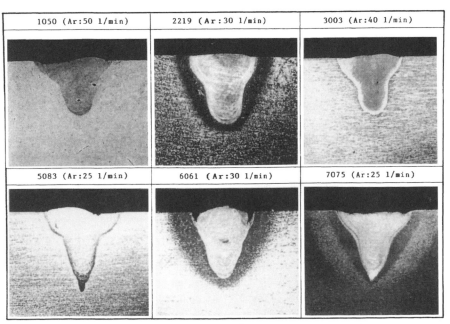

Beam Power : 4 kW, Welding Speed : 1 m/min, Focal Point : −4 mm

**Figure 5.15.** Examples of penetration shapes by $CO_2$ laser welding in various Al alloys. From Matsumura et al. (1992).

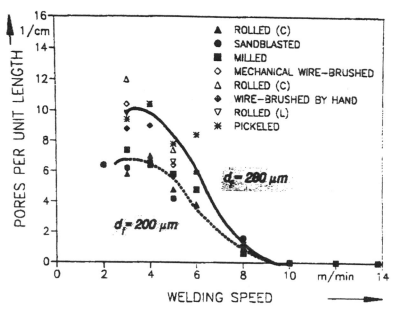

**Figure 5.16.** *Number of pores per unit length as a function of welding speed for AlMgSi1 with a plate thickness of 5 mm (P, 4.2 kW; F, 4; top bead shielding gas, pure He). From Rapp et al. (1993).*

higher welding speed and high beam intensity. This suggests that collapse of the keyhole is an important factor in the production of pores. Kim, Watanabe, and Yoshida (1993) have shown that ultrasonic vibration during Nd:YAG laser welding of 5000 and 6000 series Al alloys can result in a reduction in porosity and hot cracking under certain conditions. A gas chromatographic study of the composition of the gas within pores in laser-welded Al alloys failed to detect significant quantities of hydrogen but did find that the shield gas was present (Simidzu et al. 1992). It was suggested that vaporization of Mg played a dominant role in the creation of cavities. Eberle, Richter, and Schobbert (1994) have shown that pore-free welds also can be produced through optimized modulation of the laser beam. It seems this technique may act to stabilize the keyhole through coupling to the fundamental mode structure in the liquid sheath surrounding the keyhole.

The mixing of alloy components in the weldment and on either side of butt welds between dissimilar Al alloys was investigated by Gu (1995). The distribution of Cu, Si, and Zn across such a weld between 7075 and 6061 alloys is shown in Figure 5.17. The $CO_2$ laser beam was scanned along the seam between the two sheets (region 2) such that an equal area on either side of the seam was exposed to laser radiation. Regions 1 and 3 correspond to the HAZ on either side of the fusion zone. Diffusion of alloying elements into and out of the parent alloys at the boundaries of these regions is clearly indicated in this fusion weld.

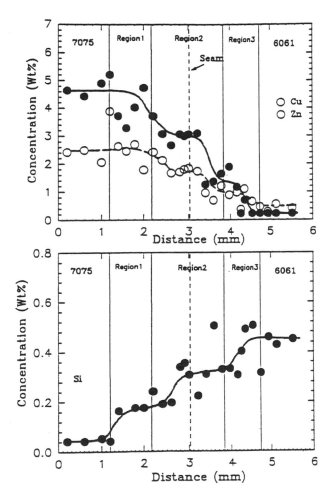

**Figure 5.17.** *Element distribution of Zn, Cu, and Si across the weldment for a butt weld between 7075 and 6061 alloys. Solid lines, boundaries of the weld pool and laser beam irradiation zone. Dashed line, seam and laser beam center. From Gu (1995).*

## 5.3.4 Other Alloys

A summary of microstructural studies in the fusion zone of other laser-welded alloys is given in Table 5.3.

**TABLE 5.3. Microstructural Studies of Welds in a Variety of Alloys**

| Alloy* | Welding Condition | Note | Reference |
|---|---|---|---|
| NAB | 2–13.8 kJ/cm $CO_2$ | | Bell, Petrolonis, and Howell (1992) |
| | 4–14 kJ/cm $CO_2$ | | Petrolonis (1993) |
| Ni-Al-Cr, Mg-Zn-Zr | | Corrosion behavior | Kattamis (1981) |
| Ti-Al-Nb | 5 kJ/cm Nd:YAG | Postweld treatment | Baeslack, Cieslak, and Headley (1989) |
| Ti6A14V-Beta-C | 0.7 kJ/cm $CO_2$ | Compare GTA | Liu, Baeslack, and Hurley (1994) |
| Ti22V4Al | 0.2–0.6 kJ/cm $CO_2$ | | Shinoda, Matsunaga, and Akaishi (1992) |
| Cu-steel | | | Dell'Erba et al. (1986) |
| Cu-SS304 | 1.2 kJ/cm $CO_2$ | X-ray map | Gopinathan et al. (1993) |
| Ti-Al-Mo | 0.5 kJ/cm | Preheat | Hirose, Arata, and Kobayashi (1995) |

# 6

# *Welding of Nonmetals*

## 6.1 INTRODUCTION

The strong interest in and emphasis placed on laser welding of metals should not be allowed to obscure the fact that laser radiation is potentially useful in the welding of other materials. With such a strong focus on welding metals, there has been little work done on laser welding of other important industrial materials such as ceramics, metal matrix composites (MMCs), and polymers. This chapter reviews data on laser welding of these materials.

## 6.2 LASER JOINING OF CERAMICS AND GLASSES

$CO_2$ and Nd:YAG laser radiation is strongly absorbed by silicate and oxide materials, with absorption occurring over a depth of $10-100$ $\mu$m. This leads to strong surface melting in ceramics such as $Al_2O_3$, $Y_2O_3$, and $ZrO_2$ at relatively modest laser powers. Rapid vaporization at the laser focus can result in keyhole formation as in metals, with the possibility of penetration or keyhole welding conditions. Although this efficient use of laser radiation is attractive and alternative ceramic joining techniques are of great interest, several limitations exist that must be taken into account.

low thermal shock resistance, leading to cracking

heating and vaporization lead to porosity in the weld region and heat-affected zone (HAZ)

a tendency for grain growth in both the fusion zone and HAZ
solidification must be controlled to promote the desired recrystallization

In practice, preheating of the ceramic before laser welding has been shown to
eliminate many of these problems (Maruo, Miyamoto, and Arata 1985, Tönshoff
and Gonschior 1993, Exner et al. 1993), so laser welding has become an attractive
alternative technique for joining oxide ceramics.

$CO_2$ laser welding of sintered $ZrO_2$ stabilized with CaO, MgO, or $Y_2O_3$ was
reported by Maruo, Miyamoto, and Arata (1985) and followed an earlier study of
$CO_2$ laser welding of $SiO_2$-$Al_2O_3$ ceramic (Maruo, Miyamoto, and Arata 1981). The
laser power was 1 kW, and the ceramic workpiece was placed in a furnace in air
at temperatures of up to 1400°C. Typical welding speeds were 80–120 cm/min
(Figure 6.1). Macroscopic crack formation was evident at preheat temperatures up
to ~1200°C, but these appeared only outside the fusion zone and arose due to the
high thermal gradient between the weld zone and base material. Preheating to temperatures in excess of 1200°C eliminated these macroscopic cracks, but microscopic
intergranular cracks were still observed with some compositions in the fusion zone
even after this treatment; these were attributed to shrinkage of the weld bead after
solidification. A CaO-stabilized $ZrO_2$ material with 0.7% $Al_2O_3$, 1.5% $SiO_2$, and
1.0% MgO did not exhibit these microcracks. The microstructure through the weld
zone in this material is shown in Figure 6.2. The development of a columnar growth
structure along the thermal gradient at the fusion zone–HAZ boundary is apparent,
as is the equiaxed structure toward the center of the fusion zone. The hardness was
found to increase in the region of columnar growth in which porosity was minimal
and then decrease in the region of larger grain size in the center of the fusion zone.

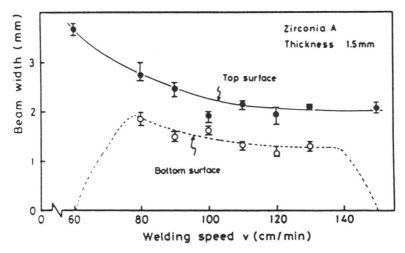

**Figure 6.1.** Effect of welding speed on bead widths at 1-kW power. From Maruo, Miyamoto,
and Arata (1985).

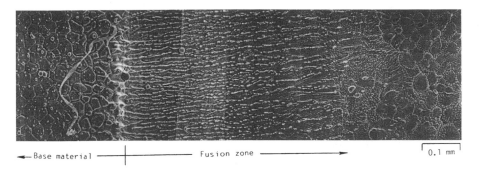

← Base material ———————— Fusion zone ————————→        0.1 mm

**Figure 6.2.** *Laser-welded bead surface of CaO-stabilized $ZrO_2$ ceramic at 120 cm/min for 1-kW $CO_2$ laser. From Maruo, Miyamoto, and Arata (1985).*

The bending strength was found to increase toward that of the base material at high welding speed (120 cm/min) but was always less than that of unwelded ceramic. Samples welded under conduction limited conditions (no keyhole) were found to have a bending strength that was somewhat larger than that of the base material.

Good-quality butt welds in $Al_2O_3$ and $Al_2O_3$-$ZrO_2$ ceramics can be produced using pulsed Nd:YAG laser heating (Exner et al. 1993, Tönshoff and Gonschior 1993) if preheating is used. An example of butt welds in $Al_2O_3$-$ZrO_2$ ceramic at various welding speeds is shown in Figure 6.3.

A thermal analysis of heating and cooling during laser welding of ceramics and the stress fields produced was conducted by Tönshoff and Byun (1992). The stress is found to be mainly compressive, but a tensional stress is predicted near the front of the weld pool and at points near the keyhole.

Laser welding of fused quartz and other glasses has been shown to be a viable technique (Phitzer and Turner 1968, Deminet 1987), suitable for attaching optical windows to tubing and other components.

A joining technique based on laser-activated brazing of $Si_3N_4$ ceramic offers an alternative approach to laser welding (Miyamoto et al. 1987). This technique uses laser radiation to heat a solder powder placed on top of the two plates to be butt welded. Another laser beam irradiates the bottom of the plates. When the solder powder melts, it is drawn into the seam to form the bond. The laser power is then reduced in a programmed sequence to minimize thermal stresses in the two plates. An acceptable bond strength was reported by Miyamoto et al. (1987).

## 6.3 HETEROGENEOUS COMPOSITIONS

MMCs are used in a variety of aerospace applications as well as in automotive manufacturing. They combine high strength with ductility and have excellent dimensional stability plus low weight. The joining of MMCs is a necessary but challenging task in view of their tendency to form brittle phases on heating in response to a

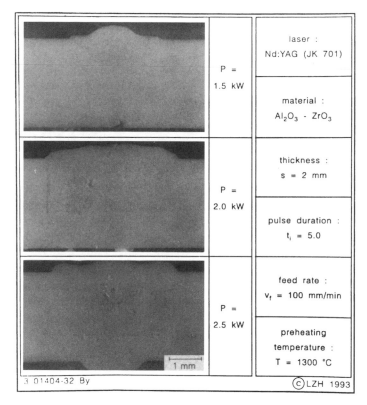

**Figure 6.3.** *Butt welds in Al₂O₃-ZrO₂ ceramic using pulsed Nd:YAG laser radiation. From Töns-hoff and Gonschior (1993).*

reaction between the matrix alloy and the reinforcing phase. Crack initiation in these phases results in brittle fracture and poor weld strength.

The control over heating and cooling rates possible with laser radiation has led to a number of welding trials with several MMC materials. $CO_2$ and Nd:YAG laser sources in both continuous-wave (CW) and pulsed mode have been used. Good results are obtained under certain conditions, but laser welding has not yet been proved to be a reliable method of joining these materials.

Most experimental work on laser welding of MMCs has been carried out with SiC-Al composites (Dahotre, McCay, and McCay 1989, 1990, 1991, McCay, McCay, and Dahotre 1991, Gopinathan, Murthy, and McCay 1993). Typical compositions were 10–20 vol.% SiC particulates in A356 Al alloy. This alloy has 7.0 wt.% Si, 0.35 wt.% Mg, <0.11 wt.% Zn, <0.2 wt.% Fe, and <wt.% Cu. The SiC is present as ~10-$\mu$m particles in intercellular Al regions.

Laser heating has several effects on composition and structure due to the heterogeneous nature of this material. Because SiC particles are strongly absorbant at $CO_2$ and Nd:YAG wavelengths, direct exposure to incident laser radiation on, for example, the walls of the keyhole, will result in vaporization of SiC, liberating both Si and C into the plasma and weld pool. The average size of SiC particles remaining after exposure to laser radiation also will be reduced, and their morphology and composition will be altered. An important reaction involves the formation of carbides such as $Al_4C_3$ via reaction with Al:

$$4Al_{(e)} + 3SiC_{(s)} \rightarrow Al_4C_{3(s)} + 3Si_{(s)}$$

or directly through the reaction of carbon atoms liberated in the laser-induced decomposition of SiC particles. The microstructure of SiC/A356-Al MMC before and after laser melting is shown in Figure 6.4 (Gopinathan, Murthy, and McCay 1993). The central fusion zone (region A) consists of plate-like $Al_4C_3$ needles in a fine dendrite matrix. Region B, surrounding the central fusion zone, shows redistributed SiC particles whose edges have been melted and surfaces modified through exposure to laser heating. The partially melted zone C contains a fine cellular/dendritic structure.

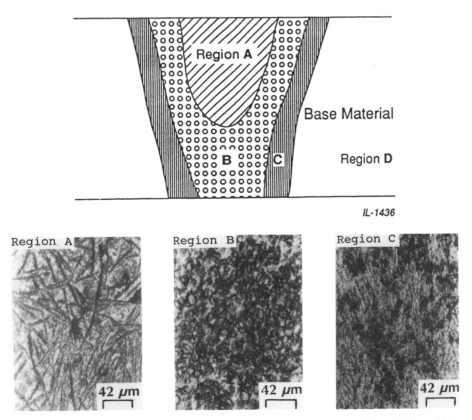

IL-1436

**Figure 6.4.** (Top) Schematic of the laser-processed microstructure in SiC/Al MMC: A, $Al_4C_3$ region; B, SiC redistribution region; C, Partially melted region; D, Unaffected base MMC region. (Bottom) microstructure in regions. From Gopinathan et al. (1993).

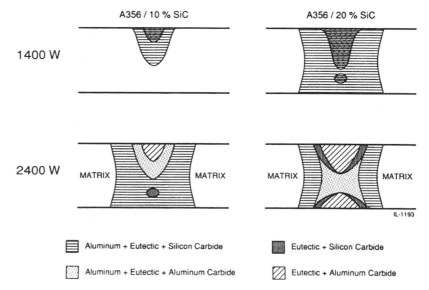

**Figure 6.5.** Microstructural composition of weld region in SiC-Al356 composite at different laser powers and MMC compositions for $CO_2$ laser at 2.5-cm/sec weld speed. From McCay et al. (1991).

The evolution of these different regions in and near the weld zone has been found to depend on SiC concentration and on laser power at a constant welding speed (McCay et al. 1991). A schematic of weld structure versus processing power and MMC composition is shown in Figure 6.5.

The microhardness in the fusion zone was found to increase substantially over that of the base material (Dahotre, McCay, and McCay 1989). This change occurred smoothly with distance through the HAZ from the base material. The presence of precipitates along dislocations was suggested as the source of this additional hardness. The sensitivity of these results to thermal history was investigated by Dahotre, McCay, and McCay (1991) using pulsed laser radiation with a variable duty cycle. A comparison of tensile test data showed that processing with pulse duty cycles between 67% and 74% (Table 6.1) yielded a tensile strength and elongation in fully penetrated laser-welded samples that were comparable to those of the parent material.

In a related study on $CO_2$ laser welding of an alumina-reinforced 6061 aluminum alloy, Kawali, Vieglahn, and Scheuerman (1991) found that the laser plasma was an important factor. Explosive expansion of this plasma was identified as a cause of poor weld performance and the generation of pores. The welds produced in this series of experiments were of poor quality.

The use of a laser-activated filler material to promote bonding in MMCs was discussed by Dahotre, McCay, and McCay (1993). This technique uses Ti or a Ti alloy as a reactive filler to form TiC instead of the $Al_4C_3$ formed through reaction of Al with SiC. This was shown to prevent the formation of plate-like $Al_4C_3$ struc-

**TABLE 6.1. Tensile Test Data for Autogenous Laser-Joined SiC Particulate/Al Alloy Composite**

| Material | Duty Cycle (%) | Elastic Modulus (GPa) | Ultimate Tensile Stress (MPa) | Elongation (%) |
|---|---|---|---|---|
| 20vol. % SiC/Al | As received | 23.00 | 302 | 2.3 |
| | 67 | 22.00 | 358 | 2.6 |
| | 74 | 21.00 | 329 | 2.6 |
| 10vol. % SiC/Al | As received | 20.00 | 262 | 3.2 |
| | 67 | 20.00 | 300 | 3.0 |
| | 74 | 20.00 | 304 | 3.7 |

From Dahotre, McCay, and McCay (1993).

tures. The role of the laser in this bonding technique is to induce a reaction between Ti powder applied at the interface between two SiC-Al sheets and to heat the base material to minimize thermal stresses.

## 6.4  POLYMERS

$CO_2$ laser radiation is readily absorbed by most polymers, but the attenuation length for absorption depends on the strength of infrared absorption bands that overlap with the wavelength emitted by the laser. In the absence of any wavelength selection via tuning of the laser cavity, the output of a $CO_2$ laser operating in CW or pulsed mode occurs sequentially on the P(18), P(20), and P(22) lines of the 10.4-$\mu$m band (Duley 1976) at wavelengths of 10.57, 10.59, and 10.61 $\mu$m, respectively. By incorporating a wavelength-selective element in the laser cavity, the output wavelength can be tuned over a range of ~9.12 to ~11.0 $\mu$m, although the laser gain and, thus, output power are reduced from that obtained at ~10.6 $\mu$m. In principle, this enables the laser to be adjusted for optimum penetration in a specific material, but absorption of polymers at these wavelengths often is weak. This results in a large penetration depth and creates an extended heat source within the material. In the case of polypropylene tuning, the laser wavelength to 10.3 $\mu$m (970 cm$^{-1}$) would dramatically increase absorption and diminish penetration depth because this wavelength is preferentially absorbed by this polymer.

The welding of thermoplastics with $CO_2$ laser radiation was first reported in 1972 (Ruffler and Gürs 1972, Duley and Gonsalves 1972). An example of a high-quality weld in polyethylene sheet is given in Duley (1976). Subsequent data on $CO_2$ laser welding of polymer are available from Duley and Mueller (1992). Under optimized conditions, laser irradiation can initiate melting and the formation of a melt pool without decomposition. The temperature range between the melting point and that at which thermal decomposition occurs is small, however, in thermoplastics. For example, in polypropylene, the melting temperature, $T_m$, is ~327°C, whereas decomposition in an inert atmosphere occurs at $T_d$ ~ 387°C. In air, the decomposition temperature, defined as that at which the polymer loses half its weight after 40 minutes, is $T_d \sim 250°C$. Laser welding of plastics then represents a delicate balance between melting and decomposition. In polymethyl methacrylate, the solid decomposes directly (sublimation) without passing through a liquid phase. As a result, laser welding of polymethyl methacrylate is not possible. Thermoset plastics, such as phenolics and epoxies, cannot be welded.

With thermal degradation, the mechanical properties of the polymer are dramatically altered; these include mechanical strength, stiffness, and elongation. Cross-linking in thermoplastics to form small solid masses that are resistant to melting can occur when the thermal degradation temperature is reached. These masses can compromise the mechanical integrity of the heated material. Overall, these constraints lead to a limited working range for laser welding of polymers (Nonhof 1994), although if conduction welding conditions can be established without a rise in surface temperature into the range of $T_d$, then good-quality welds can be generated. Because the thermal conductivity, K, of thermoplastic polymers is small, with K ~ 0.1–1 W/m°C, and the thermal diffusivity typically is $\kappa \sim 10^{-7}–10^{-6}$ m²/sec, this can be accomplished only in thin samples under surface heating conduction welding conditions. An alternative may exist in some polymers, however, when the absorption and scattering coefficient for laser radiation are small. In this case, which can occur in polyethylene materials, heat is distributed over a volume $\sim A\alpha^{-1}$, where A is the area of laser focus on the surface and $\alpha$ (cm⁻¹) is the absorption/scattering coefficient at the laser wavelength. When this occurs, the laser radiation may be considered to form a linear heat source of depth $\alpha^{-1}$ cm. For a sheet of finite thickness, heat transfer is described approximately by the Swift-Hook and Gick (1973) formalism with the following solution:

$$\frac{vW}{\kappa} \sim \frac{0.48 \, P}{dKT}, \tag{6.1}$$

which is valid at high welding speed. With $\kappa = 10^{-7}$ m²/sec, $K = 0.5$ W/m°C, and $T = 200°C$,

$$v = \frac{5 \times 10^{-10} \, P}{Wd} \text{ (m/sec)}, \tag{6.2}$$

and for $W = d = 1$ mm, $v = 5 \times 10^{-4} P = 0.5 P$ (mm/sec), where $P$ is given in watts. This illustrates the potentially efficient nature of laser welding in polymers; a laser power of ~10 W is sufficient to weld 1-mm stock at 5 mm/sec.

Laser welding is a promising yet unexploited technique for the joining of certain thermoplastics. Advantages seem to lie in the joining of thicker stock (>1 mm) in materials such as polyethylene.

# 7

# *Techniques for Improving Welding Efficiency*

## 7.1 GENERAL

The clean, undisturbed surface of metals is highly reflecting at both $CO_2$ and Nd: YAG laser wavelengths, but this reflectivity diminishes with an increase in temperature. Many metals also rapidly oxidize when heated in air to temperatures near their melting point. The result is an overall increase in absorption of laser light that depends on the product $I\tau$, where I is incident laser intensity, and $\tau$ is the dwell time at a particular point on the metal surface. (Kinsman and Duley 1986)

A thermal runaway effect at large values of $I\tau$ leads to vaporization, surface disruption, and the establishment of a keyhole. An example of this effect is shown in Figure 7.1 for Nd:YAG and $CO_2$ laser welding of an Al alloy. The role of laser wavelength is clearly evident in the lower-intensity threshold for welding with Nd: YAG laser radiation.

The welding efficiency defined as a power (or energy) transfer coefficient, $\eta$, where

$$\eta = \frac{\text{Laser power absorbed by workpiece}}{\text{Incident laser power}}, \tag{7.1}$$

is small below the threshold for keyhole formation but can approach unity once a keyhole has been established. The melting efficiency or melting ratio, $\epsilon$, where

$$\epsilon = \frac{\mathrm{v}dW\Delta H_m}{P}, \tag{7.2}$$

relates the rate of melting to incident laser power, $P$, where v is welding speed, $d$

**145**

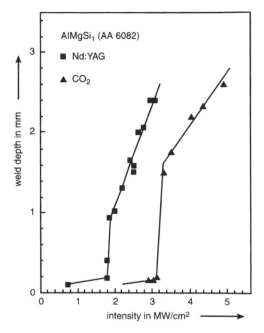

**Figure 7.1.** *Weld depth versus laser intensity for Nd:YAG and CO₂ laser radiation. From Dausinger et al. (1996).*

is sheet thickness, $W$ is beam width, and $\Delta H_m$ is the heat content of the metal at the melt temperature. The maximum value for $\epsilon$ is 0.48 for penetration welds and 0.37 for conduction welds (Swift-Hook and Gick 1973). It is apparent that $\epsilon$ never approaches unity even when $\eta \cong 1$.

Below the threshold for keyhole formation, both $\eta$ and $\epsilon$ can be enhanced if the absorption coefficient, A, can be increased. This can be accomplished in a number of ways (Duley 1986):

1. application of an absorbant coating
2. surface roughening or texturing
3. preheating
4. tailoring of temporal irradiation profile
5. oxidation/nitridation

## 7.2  ABSORBANT COATINGS

These coatings were discussed by Steen (1986) and Steen, Chen, and West (1987). The properties of some coatings are summarized in Table 7.1. Desired characteristics of these coatings are:

**TABLE 7.1. Absorbant Coatings for Laser Surface Melting**

| Coating | Comments | Absorption (%) |
|---|---|---|
| Colloidal graphite | May sublime or burn off before melting occurs with consequent falling in the coupling; there will be some carbon alloying from the coating if melting is achieved. | ~78 |
| Manganese dioxide | Decomposes to give off oxygen | |
| Manganese phosphate | A byproduct of some casting processes; there is a chance of some of the phosphorus entering the melt pool. | ~75 |
| Zinc phosphate | Same as for manganese phosphate | |
| Black paint | Paint may disappear before melting occurs | ~95 |
| Alkaline halides, sodium or potassium silicate (water glass) + graphite | Does not fail on heating but may be alloyed into the melt pool; "graphitized slag" of this form has been found to be superior to phosphatizing. | ~80 |
| Sandblasting | Survives heating and involves no alloy addition | ~70 |

From Steen, Chen, and West (1987).

1. ease of application

2. adhesion to metal surface

3. good thermal conduction to metal surface

4. chemical stability

5. no interference with weld metallurgy

Black paint satisfies many of these criteria but can be lost during welding due to volatilization of the binder in the early stages of heating, which reduces the effective value of $\eta$. Colloidal graphite is more stable, but alloying of carbon is a common side effect.

## 7.3 SURFACE ROUGHENING AND TEXTURING

Sandblasting or rubbing with sandpaper is effective in increasing the surface roughness of metals, with the scale of roughness dependent on grit size, mechanical pressure, gas flow, and so on. Figure 7.2 shows this effect for different treatments and compares the resulting reflectivity with that obtained through painting and coating with various materials. The largest reduction in reflectivity occurs as expected after sandblasting with grit that has a size comparable to that of the laser wavelength (10.6 $\mu$m).

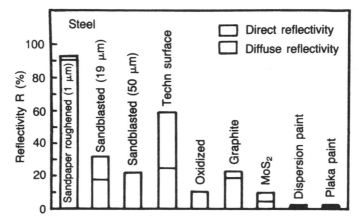

**Figure 7.2.** *Reflectivity of steel surface at 10.6 μm after various surface treatments. From Steen (1986).*

Although an impressive reduction in R occurs with sandblasting, the application of this procedure in a productive environment has many obvious limitations. As a result, sandblasting is not a practical alternative to other techniques for increasing A. Similar limitations apply to the use of grit paper. A comprehensive study of the effect of roughening with SiC paper on the absorptivity of the surface of 304 stainless steel was made by Weiting and de Rosa (1979). It was found that although an increase in A was present at $T < 400°C$, this enhancement decreased near 600°C as the defects created by roughening were annealed out.

The anodization of Al before laser welding has been shown to increase weld penetration as well as melt volume (Mallory, Orr, and Wells 1988). The effect of different surface finishes on A and weld properties is summarized in Table 7.2. All welds were porous, but the degree of porosity was not correlated to surface finish. A similar result was obtained by Duley and Mao (1994).

Surface roughening and texturing with excimer laser radiation before welding have been shown to be effective in increasing laser coupling efficiency at intensities near threshold (Duley, Mao, and Kinsman, 1991, Duley and Mao 1994). The depen-

**TABLE 7.2. Effect of Surface Finish on Weld Geometry and Beam Absorption**

| Surface Finish | Average Depth (mm) | Average Width (mm) | Average Absorption Coefficient |
|---|---|---|---|
| C  Clear sulfuric acid anodized | 1.856 | 2.225 | 0.542 |
| B  Black sulfuric acid anodized | 1.802 | 2.215 | 0.481 |
| P  Phosphoric acid anodized | 1.120 | 1.421 | 0.268 |
| A  As rolled | 0.770 | 1.135 | 0.154 |
| E  Caustic etched | 0.465 | 0.700 | 0.169 |

From Mallory, Orr, and Wells (1988).

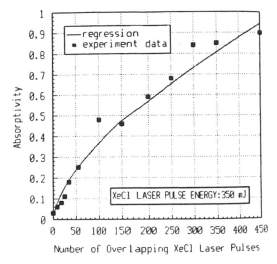

**Figure 7.3.** $\alpha'$ (10.6 μm) = A versus number of pulses for excimer laser–irradiated Al 1100. From Duley, Mao, and Kinsman (1991).

dence of the absorption coefficient, A, at a wavelength of 10.6 μm on the number of overlapping XeCl laser pulses is shown in Figure 7.3. The peak XeCl laser intensity was ~$10^9$ W/cm$^2$. It is apparent that A can approach unity after heavy processing; this is attributed to a combination of surface roughening and oxidation (Duley 1996). Figures 7.4 and 7.5 show the effect of excimer laser preprocessing on energy transfer

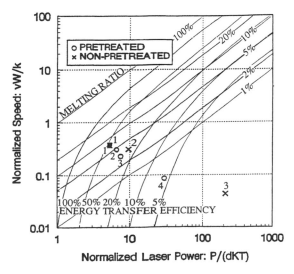

**Figure 7.4.** *Energy transfer efficiency and melting ratio for Al 1100 under pretreated and pristine conditions. From Mao (1993).*

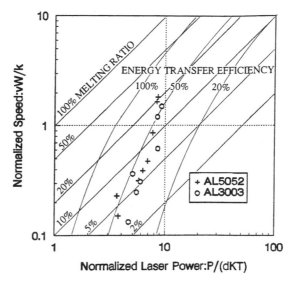

**Figure 7.5.** *Theoretical and experimental energy transfer efficiency and melting ratio in the $CO_2$ laser welding of Al 3003 and Al 5052 with excimer laser pretreatment. From Mao (1993).*

efficiency ($\eta$) and melting ratio ($\epsilon$) for Al 1100, Al 5052, and Al 3003 for laser welding at 10.6 $\mu$m (Mao 1993). A comparison between welding of pristine and pretreated Al 1100 shows that the dominant influence of excimer laser–induced surfacing occurs at low laser power (i.e., near threshold). Melting ratios approaching 20% are obtained in excimer laser–treated Al 5052 and 3003 (Figure 7.6). Similar

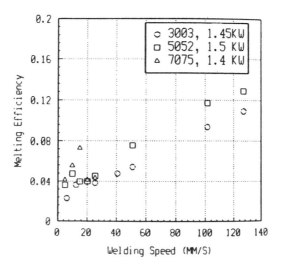

**Figure 7.6.** *Melting efficiency versus welding speed for Al 3003, Al 5052, and Al 7075. From Duley, Mao, and Kinsman (1991).*

effects were observed in other alloys, including Al 7075, although weld quality was not generally improved through pretreatment with excimer laser radiation.

Additional studies on the effect of laser-induced surface structures on absorption and coupling of 10.6-$\mu$m $CO_2$ laser radiation into metals were conducted by Ursu et al. (1984, 1987). The role of surface periodic structures that can be created on metal surfaces through irradiation with short-duration high-intensity pulses was been discussed in Section 4.4. Surface rippling also may occur as a dynamic process during CW irradiation and can contribute to the onset of keyhole formation.

A related application involves the technique developed by Williams et al. (1993) for the removal of the Zn coating from galvanized steel before $CO_2$ or Nd:YAG laser welding. This has the advantage that the elimination of Zn before welding can lead to a more stable welding process, with improvements in weld quality and increased penetration. Williams et al. (1993) concluded that complete removal of the Zn coating may not be necessary to obtain an improvement in welding, with the primary effect of excimer pretreatment being the creation of micropores that permit Zn vapor to escape during welding.

The concept by which pulses from an excimer or Nd:YAG laser are superimposed on the laser focus during welding with CW $CO_2$ laser radiation was developed by O'Neill and Steen (1988). Under certain pulse energy and irradiation conditions, an enhancement in welding speed was observed, but the overall effect seemed to be insufficiently significant to merit the extra cost and complexity associated with an additional laser. A similar caveat can be applied to excimer laser pretreatment of metals before welding.

## 7.4   BEAM POLARIZATION

The importance of beam polarization in laser cutting and welding has been recognized for some time (Arzuov et al. 1979, Olsen 1980, Beyer, Behler, and Herziger 1988). This effect occurs because of the strong dependence of the reflection coefficient at oblique incidence on the direction of the electric field vector relative to the plane of incidence. Oblique incidence occurs naturally in laser welding when a keyhole is present as the incident laser beam impinges on the leading edge. If the polarization vector is in the plane of incidence (p-polarization), then strong absorption will occur. This enhances energy deposition into the leading edge of the keyhole and results in increase weld penetration.

The original experiments of Beyer, Behler, and Herziger (1988) showed a significant effect of beam polarization on welding depth only at high weld speed in steel. Surprisingly, no effect of beam polarization was observed in the $CO_2$ laser welding of Al. Subsequent experiments (Sato, Takahashi, and Mehmetti 1996) have demonstrated that an effect exists (Figure 7.7) but that the shield gas affects the coupling efficiency and may obscure the dependence on polarization. It appears that absorption in the keyhole plasma dominates Fresnel absorption or at least is comparable to Fresnel absorption and that this process acts to obscure the effect of beam polarization.

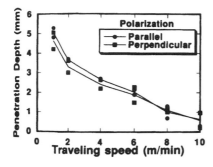 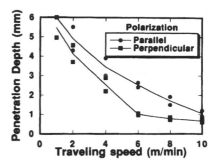

**Figure 7.7.** Dependence of penetration depth on welding speed for two laser beam polarizations in (left) Ar and (right) He shield gas. From Sato, Takahashi, and Mehmetli (1996). (Reprinted with permission from Journal of Applied Physics Copyright 1996. American Institute of Physics)

A strong polarization dependence has, however, been found in pipe welding when the laser beam is directed into the wedge formed as the two sides of the pipe are brought together to form a continuous seam (Minamida et al. 1991, Behler, Beyer, and Schafer 1988b).

## 7.5 BEAM COMBINATION AND SPLIT-BEAM TECHNIQUES

Advantages associated with the use of two laser sources in welding were discussed by Glumann et al. (1993a, 1993b) and Dausinger et al. (1995). With this method, the beams from two separate laser sources can be overlapped at the surface of the workpiece with independent control over beam polarization, temporal modulation, and spatial intensity distribution. This offers considerable flexibility with respect to control over heating and cooling rates, keyhole location, and melt pool dynamics. One laser beam also may be directed to produce preheating or postheating of the metal along the weld line. Some advantages, as identified by Glumann et al. (1993a,b) include:

improvement in the stability of the welding process

control over porosity and the formation of weld defects

improvements in weld fatigue behavior

These advantages must be balanced, however, against increased capital cost and additional complexities in beam focussing.

The system developed by Glumann et al. (1993a) for beam combination with two 5-kW $CO_2$ lasers is shown in Figure 7.8. The polarization of each beam can be independently controlled. With Nd:YAG lasers, fiber-optic delivery allows beam combination at a single welding bead (Dausinger et al. 1995).

Experimental studies of the welding of various metals with the use of this tech-

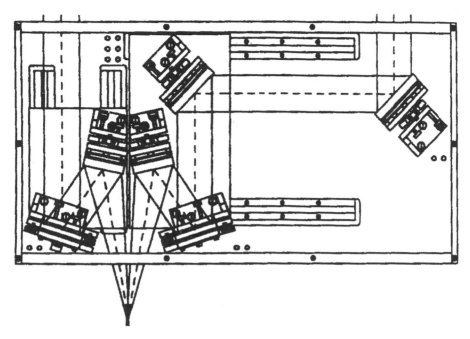

**Figure 7.8.** *Optical set-up for focussing and combination of 2 $CO_2$ laser beams. From Glumann et al. (1993a).*

nique have shown that when the two laser beams are overlapped and in focus, the penetration depth is the same as that obtained for a single laser with a power equal to the sum of that of the two individual laser sources (Glumann et al. 1993a). Subtleties in the beam combination technique are evident when the beams are adjusted to combine through the addition of angles or diameters. In the former case, a single spot occurs at the focus, whereas with the latter case, two separate spots appear. A difference in penetration depth in the welding of Al 6110 alloy with two Nd:YAG beams shows that there clearly is some advantage in adopting the "addition of diameters" technique, although the dependence of penetration depth on the method of combining beams is found to not be strong (Dausinger et al. 1995).

A split-beam technique, which involves directing a portion of a single laser beam into a different position relative to the primary focus, was investigated theoretically (Kannatey-Asibu 1989, 1991, Liu and Kannatey-Asibu 1993, Chen and Kannatey-Asibu 1994). Beam geometry is as shown in Figure 7.9, with a fraction of the main laser beam directed so as to yield preheating of the metal before welding. The overall effect of this procedure is to reduce the cooling rate and thereby reduce brittle microstructure in high-hardenability materials. The preheating beam may be defocussed into a line. The use of a single beam focussed into an oblong spot oriented along the weld direction was discussed by Mombo-Caristan (1996b) and seems to offer an advantage when welding at high speed through minimization of weld instabilities.

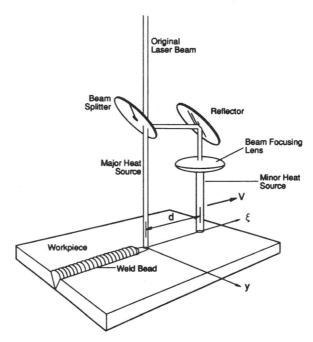

**Figure 7.9.** *Schematic of the split-beam laser welding system. From Kannatey-Asibu (1991). (Reprinted with permission ASME Journal of Engineering Materials)*

The results of these simulations show that through the judicious choice of beam separation and laser power, cooling rate can be reduced significantly over that which would be obtained with a single high-power beam. An experimental verification of these predictions would be valuable.

## 7.6   BEAM WALKING

There is a finite time delay between the onset of keyhole formation and plasma formation in which the laser beam efficiently vaporizes material. This process is interrupted by the formation and expansion of a plasma, effectively reducing the average rate of material processing. One approach developed by Arata (1987) minimizes this effect by periodically moving the laser beam ahead to dwell on a new focal area ahead of the original position of the keyhole. This simple technique combines the advantages of CW and pulsed laser welding and leads to improved penetration.

## 7.7   ARC–AUGMENTED LASER WELDING

The combination of arc and laser welding was demonstrated by Steen and Eboo (1979) and Steen (1980). In this process, an arc from a TIG torch placed near the

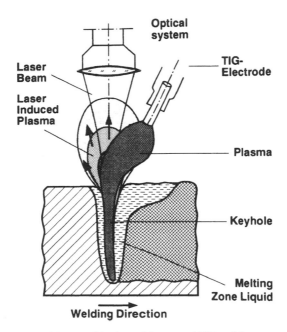

**Figure 7.10.** *Schematic of the combination of laser and TIG welding processes. From Beyer, Dilthey, Imhoff et al. (1994).*

laser focus (Figure 7.10) roots to the laser-induced keyhole when one is present or to the high-temperature zone in a conduction weld. In practice, this technique seems to be most useful in keyhole laser welding, where it offers the following advantages:

1. deposition of additional power into the keyhole, thereby simulating the effect of welding with higher laser power
2. higher welding efficiency at lower capital cost
3. additional flexibility in control over seam width and reduced sensitivity to fitup of joints
4. increased liquid melt volume, resulting in less sensitivity to welding defects
5. compatibility with both $CO_2$ and Nd:YAG laser welding sources

Rooting of the arc discharge into the laser-induced keyhole has the effect of increasing the dissipation of power within the keyhole. In general, this leads to an increase in liquid volume around the keyhole and an increase in weld width. An increase in penetration depth is obtained only after optimization of the laser focus (Matsuda et al. 1988). This suggests that electrical power from the arc is deposited relatively close to the upper part of the keyhole and produces a large increase in liquid at its entrance. Intermittent closure of the keyhole through convective motion in the liquid results in enhanced deposition of heat close to the top of the keyhole, yielding a larger weld bead without necessarily yielding increased penetration depth. High-speed welding of 12–16-mm-thick steel plate with a $CO_2$ laser/MIG source was demonstrated (Abe et al. 1996), but results were found to be dependent on gas flow and gap design. Laser plus MIG welding of a variety of Al alloys was reported by Shida, Hirokawa, and Sato (1997).

In general, the overall interaction among laser radiation, weld pool, and arc plasma is complex and means that improvements in welding conditions are obtained only after optimization of optical, geometrical, and electrical parameters. Matsuda et al. (1988) found that at a laser power of 5 kW, the addition of 300-A arc current had little effect on the weld profile when welding steel at 1 m/min, but broadening of the head of the weld bead and an enhancement of the nailhead structure were found to be important under conditions when arc augmentation is significant (Figure 7.11).

The source of this complex interaction between the laser beam and the arc when $CO_2$ laser radiation is used may arise from plasma shielding. This is less of a problem with Nd:YAG laser radiation (Beyer, Imhoff, and Neuenhahn 1994, Beyer, Brenner, and Poprawe, 1996), and the welding of steel sheets with high efficiency (40–50%) has been demonstrated (Table 7.3). Beyer, Brenner, and Poprawe (1996) discussed the use of this technique in the welding of tailor blanks.

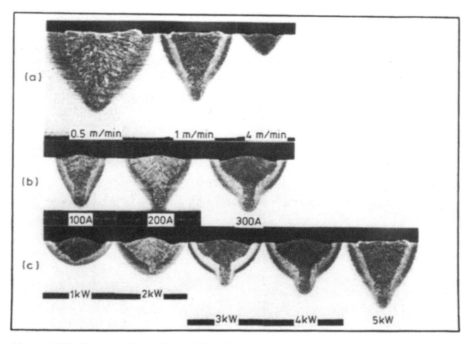

**Figure 7.11.** *Cross sections of laser TIG welds showing the effect of (a) welding speed of 0.5, 1, and 4 m/min at 5-kW laser power and 300-A TIG arc current; (b) TIG arc current of 100, 200, and 300 A at 4-kW laser power and 1 m/min welding speed; and (c) laser power of 1, 2, 3, 4, and 5 kW at 300-A TIG arc current and 1 m/min welding speed. From Matsuda et al. (1988). (Copyright TWI. Reprinted with permission)*

**TABLE 7.3. Comparison of Performance for Combined Nd:YAG/TIG Welding of Deep Drawing Steel**

| Welding Source | Power (kW) | Efficiency (%) | Speed (m/min) | Edge Preparation |
|---|---|---|---|---|
| Laser | 2 | 1–2 | 3 | Yes |
| Laser and TIG | 2 + 2 | 40–50 | 5 | No |
| Laser | 4 | 1–2 | 6 | Yes |

From Beyer, Imhoff, and Neuenhahn (1994).

Arc augmentation also is effective when laser radiation and the arc are incident from opposite sides of the workpiece. Kosuge et al. (1986) found that the optimal position for the weld electrode is several millimeters from the exit point of the keyhole. Under these conditions, the arc roots to the exit of the keyhole, resulting in an increase in melt volume at this point. The overall effect of arc augmentation is to provide a broader bead at the bottom of the keyhole and increase the stability of the keyhole. This results in reduced porosity.

Magee, Merchant, and Hyatt (1990) reported on the effect of arc augmentation on penetration depth, weld hardness, and weld microstructure in CSA 640.21/M–grade 300-W steel plate when welding with a 5.5-kW $CO_2$ laser. The arc was incident from the laser side of the workpiece. A plot of penetration depth versus total heat input (kJ/mm) (Figure 7.12) shows increased penetration over arc

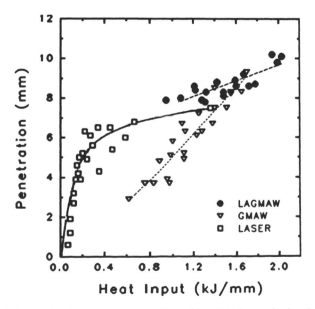

**Figure 7.12.** *Effect of heat input on penetration achieved in Laser Assisted Gas Metal Arc Welding (LAGMAW) versus Gas Metal Arc Welding (GMAW) and laser welds. From Magee, Merchant, and Hyatt (1990).*

welding alone at low heat input. For example, a penetration depth of 8 mm is obtained at 1 kJ/mm, ~50% less than that necessary for arc welding alone.

Microstructure in the HAZ was found to reflect the total heat input, with bainite and austentite the dominant phases at the HAZ–fusion zone boundary. It was found that once heat input was taken into account, there was little difference in microstructure attributable to the combined effect of arc and laser heating.

Arc augmentation of laser welding remains a potentially useful technique; however, the availability of reliable high-power $CO_2$ and Nd:YAG lasers plus the sensitivity of the arc augmentation process to alignment, electrode position, arc current, and so on may limit the applicability of this technique to certain specialized welding applications. Induction heating combined with laser welding (Beyer, Brenner, and Poprawe 1996) may yield a useful alternative and is well suited to the welding of high-carbon or high-alloy steels through minimization of the quenching rate.

# 8

## Diagnostics in Laser Welding

### 8.1  INTRODUCTION

A plume commonly accompanies laser welding under other than conduction-limited conditions. This plume is the result of the ejection of material from the area of the weld and is due to laser heating. In an ideal welding system, no vaporization would be produced, even when the surface of the material has been disturbed in response to melting at the focus of the laser beam. In practice, some vaporization always occurs at the laser intensities required for welding. This vaporized material is ejected at a thermal velocity and moves through the incident beam before leaving the focal area. While vaporized material is in the laser beam, it may be heated to temperatures greatly in excess of the vaporization temperature. This heating arises, in large part, through collisions with energetic electrons. Under certain conditions, the overall effect is to produce a rapid increase in the level of ionization within the plume with the formation of a plasma.

The conversion of a plume into a plasma can occur only if a mechanism exists for energy deposition on a time scale that is fast compared with the expansion time of the plume. In general, this cannot occur via direct absorption of incident laser photons by atoms because usually no resonance exists between the photon energy and atomic transition energies. A resonance does exist, however, for transitions between continuum states of electrons in the field of a nearby ion. Such transitions are allowed for all photon energies and provide the means by which electrons may be rapidly excited to states of high kinetic energy through absorption of laser photons. This inverse Bremsstrahlung process removes energy from the incident laser beam and redirects this energy into heating of the gas. At high electron densities, inverse Bremsstrahlung leads to rapid heating and a large attenuation coefficient for incident laser radiation. Decoupling of laser radiation from the surface of the target may occur under these conditions, and this can be responsible for an interruption in the welding process.

Plasma heating by inverse Bremsstrahlung is initiated through the presence of a low density of ''seed'' electrons in the plume. These electrons are produced through thermal ionization of vaporizing atoms and through thermionic emission. The density of these electrons may be estimated from the equilibrium between ionization and recombination at the vaporization temperature and typically is many orders of magnitude smaller than the electron densities measured in laser-produced plasmas.

The location of the plasma in laser welding is time and intensity dependent. Under certain conditions, the plasma appears to evolve from inside the keyhole into the ambient medium. At a relatively low laser intensity, the plasma remains attached to the entrance of the keyhole and is not important in attenuating incident laser radiation (Figure 8.1, type 1). At higher laser intensity (type 2), the plasma separates from the surface but still is relatively stable. Further heating produces instabilities as the plasma is sufficiently heated to interrupt the laser beam. The plasma then explosively separates from the surface (Figure 8.1, type 3). At very high laser intensities, a stable plasma geometry is observed to extend back from the entrance to the keyhole toward the laser (Figure 8.1, type 4). Such plasmas are strong absorbers of incident laser radiation and can dissipate a significant fraction of laser power before it arrives at the target. Free-standing plasmas induced by $CO_2$ laser radiation in Ar were studied by Tsukamoto et al. (1996) and Rockstroh and Mazumder (1987).

It is evident that plasmas can be an integral part of welding with high power laser radiation and that they modify the coupling of laser radiation into the workpiece, often on a rapidly varying time scale. Plasmas are a potential source of welding defects such as lack of penetration, porosity, and compositional changes. The exis-

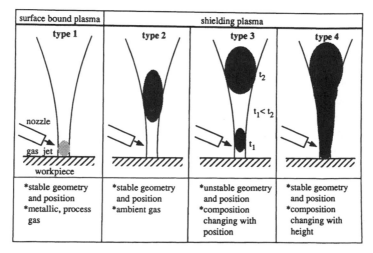

**Figure 8.1.** Type 1, plasma, low excitation, stable and in contact with workpiece. Type 2, plasma separates from workpiece but maintains stable geometry. Type 3, unstable geometry in attenuating plasma. Type 4, stable plasma, high excitation, large spatial extent. From Seidel et al. (1994).

tence of a plasma, however, provides radiation that can be sampled using optical and acoustic detectors and represents a useful monitor of welding conditions. An understanding of the relation between spectral components detected in emission from a plasma produced under these conditions and resulting weld characteristics can be obtained only with physical models.

## 8.2 PHYSICS OF LASER WELDING PLASMAS

Plasmas are characterized by their level of ionization and their excitation. The level of ionization is given by the ratio of electron density $n_e$ to the total gas density n. In plasmas above the laser focus during laser welding of metals, $n_e$ is typically $10^{15}-10^{17}$ cm$^{-3}$ at atmospheric pressure (n = $2.7 \times 10^{19}$ cm$^{-3}$). The level of ionization then is at most ~0.5%. The plasma that is formed within the keyhole during welding may have a higher degree of ionization in response to laser heating via inverse Bremsstrahlung, although measurements suggest that even under keyhole conditions, the plasma is less than completely ionized. In such weakly ionized plasmas, $n_e = n_i$, where $n_i$ is the density of singly ionized atoms. With the assumption of local thermodynamic equilibrium (LTE), the relative population of excited states can be described by a Boltzmann distribution at an excitation temperature $T_e$ that is assumed to apply to all plasma components. Under these conditions, the densities of plasma species are given by the Saha equation (Spitzer 1968):

$$\frac{n_e n_i}{n_0} = \frac{g_i g_e}{g_0} \frac{[2\pi m_e kT_e]^{3/2}}{h^3} \exp\left[\frac{-E_i}{kT_e}\right], \tag{8.1}$$

where $m_e$ is the electron mass, and $g_e$, $g_i$, and $g_0$ are the degeneracy factors for electrons, ions, and neutral atoms, respectively. $E_i$ is the ionization potential for the neutral atoms in the gas. Equation 8.1, which is subject to its limitations, can be used to relate $n_e$ and $T_e$ to the intensity of spectral lines emitted from the laser plasma. With $n_e = n_i$, equation 8.1 can be rewritten as:

$$n_e = An^{1/2}T_e^{3/4}[\exp[-\theta/T_e]]^{1/2} \text{ (cm}^{-3}). \tag{8.2}$$

Numerical values for A and $\theta$ for several elemental gases are given in Table 8.1. Both $g_i$ and $g_0$ can be large for atoms such as Fe that have many energy levels near the ground state.

**TABLE 8.1. Atomic Constants for Various Atoms and Ions and Values for A and $\theta$ in Equation 8.2: $g_e$ = 2**

| Element | $g_i$ | $g_0$ | A (cm$^{-3/2}$°K$^{-3/4}$) | $\theta$ (°K) |
|---------|-------|-------|----------------------------|---------------|
| He | 2 | 1 | $9.8 \times 10^7$ | 285,300 |
| Ar | 2 | 1 | $9.8 \times 10^7$ | 182,900 |
| Al | 1 | 6 | $2.8 \times 10^7$ | 69,400 |
| Fe | 30 | 25 | $7.6 \times 10^7$ | 91,700 |
| Zn | 2 | 1 | $9.8 \times 10^7$ | 109,000 |

With $n = 2.7 \times 10^{19}$ cm$^{-3}$, and $T_e = 10^{4\circ}$K, one obtains $n_e = 2.1 \times 10^8$ cm$^{-3}$ in He and $n_e = 4.1 \times 10^{16}$ cm$^{-3}$ in Fe, reflecting the lower ionization potential of Fe compared with that of He and the dominance of low ionization potential elements in contributing to electron density. $n_e$ versus $T_e$ for Fe and Al plasmas is shown in Figure 8.2.

Equation 8.2 also can be used to estimate the initial concentration of electrons that are the seed electrons that initiate plasma heating via inverse Bremsstrahlung. With $T_e$ replaced by $T_v$, the vaporization temperature, $n_e$, is $1.45 \times 10^8$ cm$^{-3}$ for Fe ($T_v = 3300°$K) and $n_e = 4.6 \times 10^8$ cm$^{-3}$ for Al ($T_v = 2723°$K).

The absorption and scattering of incident radiation by a weakly ionized plasma are determined by the electron density with the refractive index approximately given by:

$$n = [1 - n_e \, n_c]^{1/2} \tag{8.3}$$

where $n_c$ is a critical electron density:

$$n_c = \frac{m_e \, \epsilon_0 \, \omega^2}{e^2} \tag{8.4}$$

$$= 3.14 \times 10^{-10} \, \omega^2 \ (\text{cm}^{-3}),$$

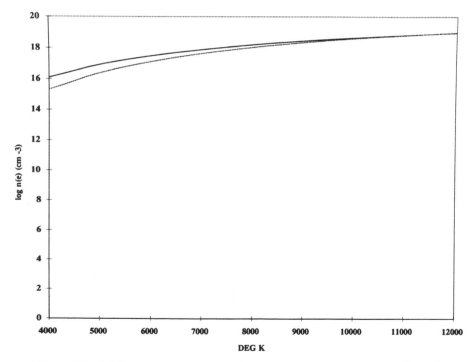

**Figure 8.2.** Solutions to equation 8.2 for Al and Fe plasma at $n = 2.7 \times 10^{19}$ cm$^{-3}$.

where $\epsilon_0$ is the permittivity of free space, and $\omega$ is angular frequency. For $CO_2$ laser radiation, $\omega = 1.78 \times 10^{14}$ rad sec$^{-1}$ and $n_c = 10^{19}$ cm$^{-3}$. When $n_e < n_c$, the refractive index is real and radiation can propagate through the plasma; it is only when $n_e = n_c$ that radiation is prevented from entering the plasma. Thus, it is evident that the reflection of 10.6- and 1.06-$\mu$m laser radiation is unimportant at the electron densities commonly observed during laser welding of metals. As a result, incident laser radiation will be transmitted through such plasmas. Absorption of this transmitted radiation by electron-ion pairs leads to plasma heating with an increase in $n_e$ and $T_e$.

The absorption and dissipation of incident laser radiation by the laser plasma can be obtained from the following time-dependent equations (Barchukov et al. 1975, Herziger, Kreutz, and Wissenbach 1986, Herziger 1986, Poueyo-Verwaerde et al. 1993):

$$\frac{dn_e}{dt} = R_i - R_d - R_r \qquad (8.5)$$

$$n_e \frac{d\bar{\epsilon}}{dt} = \alpha I - P_C - P_L, \qquad (8.6)$$

where $R_i$, $R_d$, and $R_r$ are electron production, diffusion, and recombination rates, respectively; $\bar{\epsilon}$ is average electron energy; $\alpha$ is the absorption coefficient for laser radiation; $I$ is laser intensity; $P_C$ is the power loss from elastic collisions; and $P_L$ is the power loss from inelastic collisions. The condition for rapid plasma heating is that the rate of energy input $\alpha I$ exceed the power loss due to elastic collisions. This requirement can be written as:

$$\alpha I = 2n_e \frac{m_e}{M} v \bar{\epsilon}, \qquad (8.7)$$

where $M$ is the mass of a neutral atom, and v is the collision frequency. The mean electron energy will be some fraction of $E_i$. With $\bar{\epsilon} \simeq 0.1 E_i$ (Poueyo-Verwaerde et al. 1993):

$$\alpha I \sim \frac{n_e m_e}{5M} v E_i. \qquad (8.8)$$

The inverse Bremsstrahlung absorption coefficient is (assuming LTE conditions):

$$\alpha(m^{-1}) = \frac{n_e n_i Z^2 e^6 2\pi}{6\sqrt{3}m\epsilon_0^3 ch\omega^3 m_e^2} \left[ \frac{m_e}{2\pi kT_e} \right]^{1/2} [1 - exp(- \omega/kT_e)]\bar{g}, \qquad (8.9)$$

where $Z$ is the average ionic charge in the plasma, $h$ is Planck's constant, $c$ is the speed of light, and $\bar{g}$ is the quantum mechanical Gaunt factor, which typically is 1.3–1.6 for laser welding plasmas at 10.6 $\mu$m (Dowden, Kapadia, and Postacioglu

1989, Tix and Simon 1993). For $CO_2$ or YAG laser radiation and plasmas with $T_e$ ~ 8 × $10^{3}$°K, $1 - \exp(-\hbar\omega/kT_e) \sim \hbar\omega/kT_e$, and equation 8.9 becomes:

$$\alpha(m^{-1}) = \frac{n_e n_i Z^2 e^6 \bar{g}}{6\sqrt{3} mc\epsilon_0^3 \omega^2 (2\pi)^{1/2}} \frac{1}{(m_e kT_e)^{3/2}}. \tag{8.10}$$

With $Z = 1$ (weakly ionized plasma), $\bar{g} = 1.5$, m = 1.0, and $\omega = 1.78 \times 10^{14}$ rad $\sec^{-1}$ ($CO_2$ laser radiation) equation 10 becomes

$$\alpha(m^{-1}) \sim \frac{3.3 \times 10^{-39} n_e^2}{T_e^{3/2}}, \tag{8.11}$$

where $n_e$ is given in meters$^{-3}$ and $T_e$ is given in °K. Therefore, with $n_e = 10^{23}$ m$^{-3}$ and $T_e = 10^{4}$°K, $\alpha = 33.3$ m$^{-1}$, or 0.33 cm$^{-1}$.

More detailed calculations (Szymanski and Kurzyna 1994, Finke, Kapadia, and Dowden 1990) yield similar results. Figure 8.3 shows $\alpha$ plotted versus electron temperature for mixed Fe-Ar plasmas at 1 atm pressure.

From equation 8.8, the condition for initiating an electron avalanche is

$$I > \frac{6.0 \times 10^{37} m_e \, v E_i \, T^{3/2}}{n_e M} \tag{8.12}$$

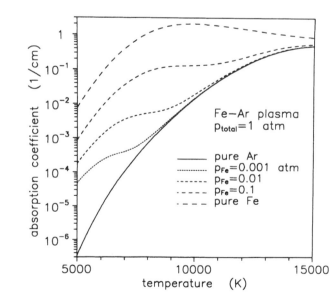

**Figure 8.3.** Absorption coefficient for 10.6-$\mu$m radiation in Fe-Ar plasmas for different Fe partial pressures. From Szymanski and Kurzyna (1994). (Reprinted with permission from Journal of Applied Physics. Copyright 1994 American Institute of Physics)

where $I$ is given in watts per meter squared and $E_i$ is given in joules. The collision frequency

$$\nu = n\sigma v \; sec^{-1}, \tag{8.13}$$

where the electron–neutral atom cross section for elastic collisions is (Tix and Simon 1993)

$$\sigma \simeq 4.25 \times 10^{-19} \; (13.6/E_e)^{1/2} \; (m^2), \tag{8.14}$$

with $E_e$, the electron energy, given in electron volts. The electron velocity $v = (2E_e/m_e)^{1/2}$ and

$$\nu_{Fe} = 9.3 \times 10^{-13} \; n \; (sec^{-1}) \tag{8.15}$$

$$= 2.5 \times 10^{13} \; sec^{-1} \tag{8.16}$$

at atmospheric pressure. Under these conditions, and with $T = 10^{4\circ}K$, $M = 55.8$ as the atomic mass for Fe, and $n_e = 10^{23}$ $m^{-3}$, the critical laser intensity $I$ is $>1.8 \times 10^7$ W/cm$^2$.

When physical parameters are known or can be estimated, equations 8.5 and 8.6 can be solved to yield the time dependence of average electron energy and electron density. An example of the results of such a calculation for a laser-induced plasma over Fe in the presence of a variety of shield gases is shown in Figure 8.4 (Herziger, Kreutz, and Wissenbach 1986). At 10.6 $\mu$m and with an incident laser intensity of $10^7$ W/cm$^2$ within a 100-$\mu$m radius focal spot, the average electron energy increases to ~1 eV over a time scale of ~2 nsec. The electron density increases at a similar rate and saturates at its equilibrium value after ~10 nsec. As expected, the equilibrium electron density is highest in Ar and lowest in He.

Power absorbed from the incident laser beam is converted to electronic excitation and therefore can be radiated at a variety of wavelengths. For a blackbody radiator, the power emitted is

$$P_{em} = 5.67 \times 10^{-8} \; T^4 \; W/m^2 \tag{8.17}$$

This is equal to $5.67 \times 10^8$ W/m$^2$ at $T = 10^{4\circ}K$. A cylindrical plasma with a radius of 100 $\mu$m and height of 1 mm, which radiated as a blackbody at this temperature, would emit ~350 W, much less than the input power. Thus, radiative losses from real laser plasmas that have emissivities much less than that of a blackbody are small and do not constitute a major cooling term. As shown later, the emission of radiation by specific atomic and ionic components within the plasma can be intense and contains useful information about plasma conditions.

For the plasma within the keyhole during laser welding, the primary energy loss mechanism arises from electron flow to the walls of the keyhole (Tix and Simon 1993, 1994). This flow occurs through presheath and sheath layers near the wall

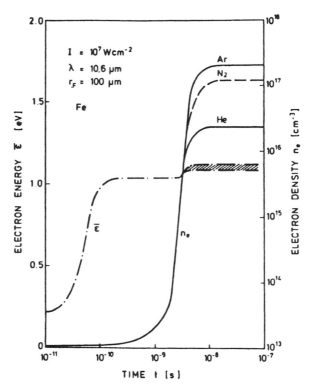

**Figure 8.4.** *Time-dependent electron density and electron energy for $CO_2$ laser radiation incident on Fe. From Herziger, Kreutz, and Wissenbach (1986).*

(Figure 8.5). The sheath is a boundary layer with a width approximately equal to that of the Debye length, $\lambda_D$, where:

$$\lambda_D = \left[ \frac{e^2(n_e kT_e + n_i kT_e)}{2\epsilon_0 \, k^2 \, T_e T_i} \right]^{-1/2} \tag{8.18}$$

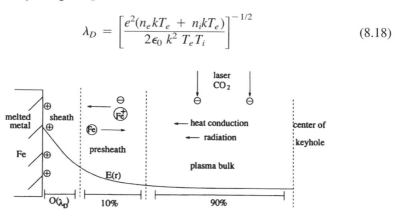

**Figure 8.5.** *Plasma structure and transport mechanisms in the keyhole. Adapted from Tix and Simon (1993).*

With $n_e = n_i = 10^{25}$ m$^{-3}$ and $T_e = T_i = 10^{4\circ}$K, $\lambda_D = 2.2 \times 10^{-8}$ m, several orders of magnitude smaller than the keyhole radius. $\lambda_D$ is a measure of the extent to which the charge of a positive ion is screened by that of electrons in the sense that the potential of an ion is not present at a distance of more than $\lambda_D$ from the ion. For distances of more than $\lambda_D$, the plasma appears to be neutral, whereas for distances of less than $\lambda_D$, charge neutrality need not be maintained and strong electric fields can be present. The electric field from this region (Figure 8.5) extends into the presheath region and acts to accelerate electrons and ions to a velocity of $\sim[k(T_e + T_i)/m_i]^{1/2}$ (Tix and Simon 1993). Electron-ion pairs recombine at the walls of the keyhole, releasing their energy as heat. The convective heat flux into the wall of the keyhole due to this mechanism dominates energy transfer from the plasma to the workpiece but amounts to only 20–30% of incident laser power. Because the radiative energy flux deposited into the walls of the keyhole from the plasma also is small, heat transfer from the keyhole plasma is not the dominant term in depositing energy into the workpiece.

Both electron and ion excitation temperatures decrease dramatically near the wall of the keyhole due to recombination, and this introduces a strong radial temperature gradient into the keyhole. Ducharme et al. (1992) discussed the optical effects that could be produced by such a gradient and concluded that refraction may occur. They found that 10.6-$\mu$m laser radiation will be refracted away from the walls of the keyhole under certain conditions. This self-focussing effect would tend to reduce Fresnel absorption at low welding speed when the keyhole radius is large. Refraction also may occur in the plasma above the keyhole, but in this case, it leads to beam defocussing (Ducharme et al. 1992).

The thermal conductivity $\epsilon\lambda$ for the partially ionized gas within the center region of the keyhole is approximately (Dowden, Kapadia, and Postacioglu 1989)

$$\epsilon\lambda = BT^{1/2} \text{ w/m/}^{\circ}\text{K,} \qquad (8.19)$$

where

$$B \sim 1.1 \times 10^{-2} \text{ W/m/}^{\circ}\text{K}^{-3/2} \qquad (8.20)$$

for iron vapor at $T_e = 10^{4\circ}$K. Then, $\epsilon\lambda \sim 1.1$ W/m/$^{\circ}$K at this temperature. Dowden, Kapadia, and Postacioglu (1989) showed that the radial temperature dependence in the center of the keyhole is:

$$T(r) \sim \{[3Q \ln (a/r)]/4\pi B + T_v^{3/2}]^{2/3} \qquad (8.21)$$

where a is the keyhole radius, $Q$ is the absorbed laser power per unit length within the keyhole, and $T_v$ is the temperature at $r = a$ (i.e., vaporization temperature). In this simple model, the average temperature within the keyhole $<T_e> \sim 1.2 \times 10^{4\circ}$K when $Q = 1.5 \times 10^3$ W/cm. This increases to $1.7 \times 10^{4\circ}$K if it is assumed that the gas within the keyhole is completely ionized (Dowden, Kapadia, and Postacioglu 1989) and the vapor consists of iron atoms. Infrared continuum emission from the laser plasma under welding conditions has been used by Ohji et al. (1995) as an indicator of electron temperature and density.

Emission of spectral lines can often be seen from the plasma during laser welding. The spectral line intensity for a transition between two states $p$ and $q$ is

$$I_{pq} = \frac{1}{4\pi} \int n_p(x) g_p A_{pq}\, h\nu_{pq}\, dx \tag{8.22}$$

where $n_p(x)$ is the density of atoms in the excited state p as a function of distance $x$ throughout the plasma, $A_{pq}$ is the atomic transition probability between the upper state $p$ and the lower level $q$, $h\nu_{pq}$ is the energy of the photon emitted, and $g_p$ is the degeneracy of the $p$ state. Calculation of $I_{pq}$ according to equation 8.22 assumes that the plasma is optically thin (i.e., no self-absorption occurs between the point at which a photon is emitted within the plasma and the surface of the plasma). This requirement can easily be met when q is an excited state. Strong self-absorption can occur, however, when q is the ground state or is within an energy $\sim kT_e$ of the ground state (Figure 8.6).

The population $n_p$ of the excited state is, assuming a Boltzmann equilibrium at temperature $T_e$,

$$n_p = \frac{n_0}{g_0}\, g_p\, \exp[-E_p/kT_e] \tag{8.23}$$

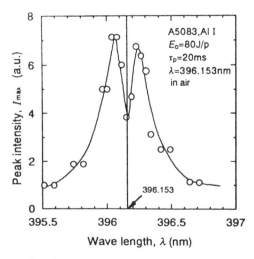

**Figure 8.6.** *Intensity profile of 396.153-nm emission line of Al in a plasma produced during welding of Al with Nd:YAG laser radiation. The dip in the center of the profile is due to self-absorption of radiation by atoms in the plasma. From Matsunawa et al. (1995).*

where $n_0$ is the population of atoms in the ground state, $g_0$ is the degeneracy of the ground state, and $E_p$ is the energy of the p state above the ground state. To estimate the plasma temperature, subject to these approximations, spectral line emission is measured at two wavelengths, $\lambda_p$ and $\lambda_s$, originating from two excited atomic states at which $E_s - E_p \gg kT_e$. Then

$$\frac{I_{sq}}{I_{pq}} = \frac{n_s}{n_p} \frac{g_s v_{sq} A_{sq}}{g_p v_{pq} A_{pq}},$$  (8.24)

so that, using equation 8.23

$$kT_e = \frac{0.434\,(E_s - E_p)}{\log\left[\dfrac{g_s A_{sq}}{g_p A_{pq}}\right] - \log\left[\dfrac{\lambda_s}{\lambda_p}\right] - \log\left[\dfrac{I_{sq}}{I_{pq}}\right]}$$  (8.25)

or

$$T_e\,(^\circ K) = \frac{5040\,(E_s - E_p)}{\log\left[\dfrac{g_s A_{sq}}{g_p A_{pq}}\right] - \log\left[\dfrac{\lambda_s}{\lambda_p}\right] - \log\left[\dfrac{I_{sq}}{I_{pq}}\right]}$$  (8.26)

where $E_s$ and $E_p$ are given in eigenvolts. Wavelengths and transition probabilities for some lines used to calculate $T_e$ for several elements are given in Table 8.2.

**TABLE 8.2. Wavelength, λ, Transition Probability gA, and $E_P$ for Spectral Lines Used to Calculate $T_e$ in Laser Welding Plasmas**

| Element | λ (nm) | gA ($10^6$ sec$^{-1}$) | $E_P$ (e/V) | Reference |
|---------|--------|------------------------|-------------|-----------|
| Ar | 714.7 | 1.386 | 13.28 | Zerkle and Krier |
| | 675.2 | 15.075 | 14.74 | (1994) |
| Ar$^+$ | 465.8 | 13.9 | 19.80 | |
| | 461.0 | 724.8 | 21.14 | |
| Al$^+$ | 281.6 | 383 | 11.82 | Knudtson et al. |
| | 466.3 | 159 | 13.25 | (1987) |
| | 559.3 | 1150 | 15.47 | |
| | 692.0 | 96 | 15.04 | |
| | 747.1 | 658 | 15.30 | |
| Al$^{2+}$ | 370.2 | 684 | 21.15 | |
| | 371.3 | | | |
| | 569.6 | 527 | 17.81 | |
| | 572.2 | | | |
| Cr | 367.869 | 830 | 3.46 | Matsunawa et al. |
| | 359.349 | 700 | 3.45 | (1995) |
| | 360.553 | 520 | 3.44 | |
| | 425.435 | 200 | 2.91 | |
| | 427.480 | 150 | 2.90 | |
| | 428.972 | 95 | 2.90 | |

The electron temperature also can be obtained by rewriting equation 8.22 as

$$\ln\left[\frac{I_{pq}\lambda_{pq}}{g_p A_{pq}}\right] = -\frac{E_p}{kT_e} + \ln\left[\frac{n_0 hc}{Z}\right], \tag{8.27}$$

where $Z$ is the partition function ($\simeq g_0$ unless there are other energy levels within $kT_e$ of the ground state). A plot of $\ell n \ [I_{pq} \ \lambda_{pq}/g_p \ A_{pq}]$ versus $E_p$ for transitions originating on several different states $E_p$ then yields a slope $1/T_e$. This method has been used to obtain $T_e$ in Al (Matsunawa et al. 1995) and Fe (Poueyo-Verwaerde et al. 1993) plasmas during laser welding. $T_e$ in laser-induced plasmas during Nd: YAG welding was calculated by Lacroix, Jeandel, and Boudot (1997).

The electron density in the plasma can be obtained under LTE conditions from the Saha equation (equation 8.1) using spectral lines of the neutral atom and its ion to calculate $n_o$ and $n_i$. Then

$$I_0 = \frac{n_0 g_0}{Z_0} A_0 \ h\upsilon_0 \ \exp\left[\frac{-E_0}{kT_e}\right] \tag{8.28}$$

$$I_i = \frac{n_i g_i}{Z_i} A_i \ h\upsilon_i \ \exp\left[\frac{-E_i}{kT_e}\right], \tag{8.29}$$

where $I_0$ and $I_i$ are intensities for lines of the neutral atom and ion, respectively; $Z_0$ and $Z_i$ are partition functions for neutral and ion, respectively; and $E_0$ and $E_i$ are the energies of the excited states in the neutral and ion, respectively, that emit at $\upsilon_0$ and $V_i$.

In the impact approximation (see Griem 1974, p. 97), the collision between electrons and neutral atoms results in the broadening of emission lines. This collisional broadening, or Stark broadening, leads to a line width of:

$$\Delta\lambda \sim wn_e + c_1, \tag{8.30}$$

where $w$ is the Stark broadening coefficient, and $c_1$ is a small term arising from ion impact. For the Fe line at 538.337 nm, $w \sim 0.0212 \ (n_e/10^{16})$ nm where $n_e$ is given in cm$^{-3}$ and $\Delta\lambda$ (equation 8.30) is the full width at half-maximum. It is apparent that an estimation of $n_e$ from Stark broadening requires high spectral resolution because $\Delta\lambda \sim 0.2$ nm at electron densities typical of laser-induced plasmas.

Electron temperatures and densities have been measured under a variety of laser welding conditions using spectroscopic measurements and the formalism of equations 8.26 and 8.30 (Sokolowski, Herziger, and Beyer 1989, Miyamoto and Maruo 1992, Collur and DebRoy 1989, Matsunawa 1990, Bermejo et al. 1990, Seidel, Beersick, and Beyer 1994, Szymanski and Kurzyna 1994, Poueyo-Verwaerde et al. 1993, Matsunawa et al. 1995, Lacroix, Jeandel, and Boudot 1997). The spatial dependence of these quantities may be obtained by imaging the laser plasma and selecting emission from individual subvolumes. The time dependence of $T_e$ can be obtained from equation 8.26 when the detection system records both spectral lines with sufficient temporal resolution (Miyamoto and Maruo 1992).

Many factors affect $n_e$ and $T_e$, so a comparison of the results obtained by different groups is difficult. A selection of representative data reported in the literature is given in Table 8.3. During welding of steel with $CO_2$ laser radiation, $n_e \sim 0.5–1 \times 10^{17}$ cm$^{-3}$ near the surface and $T_e \sim 6–8 \times 10^{3}°$K. There are few measurements of $n_e$ inside the keyhole, but those available (e.g., Maiwa, Miyamoto, and Mori 1995) suggest that $n_e \gtrsim 2 \times 10^{17}$ cm$^{-3}$ while $T_e$ rises to ~$1.3 \times 10^{4}°$K. Both $n_e$

**TABLE 8.3. Electron Density, $n_e$, and Temperature, $T_e$, during Welding of Metals with $CO_2$ Laser Radiation**

| Metal | Laser Power (kW) | Gas | h (nm) | $n_e$ (cm$^{-3}$) | $T_e$ (°K) | Reference |
|---|---|---|---|---|---|---|
| Steel | 2 | Ar | 0.5 | $2 \times 10^{17}$ | | Miyamoto and Maruo (1992) |
| Steel | 3 | Ar | 0.5 | $2 \times 10^{16}$ | 9,000 | Maiwa, Miyamoto, |
| | | | $-0.5$ | $2 \times 10^{17}$ | 13,000 | and Mori (1995) |
| Steel | 2.5 | He | 1.0 | $0.6–1.5 \times 10^{17}$ | | Ohji et al. (1995) |
| | 2.0 | | 1.0 | $1–1.3 \times 10^{17}$ | | |
| Steel | 2.5 | He | 0.0 | $4 \times 10^{16}$ | 8,000 | Szymanski and |
| | | | 2.0 | $1.8 \times 10^{16}$ | 6,100 | Kurzyna (1994) |
| | | Ar | 0.0 | $8.5 \times 10^{16}$ | 8,600 | |
| | | | 2.0 | $3.5 \times 10^{16}$ | 7,200 | |
| Fe | 10.5 | Ar | | $8 \times 10^{16}$ | 6,200 | Verwaerde et al. (1995) |
| Fe | 1 | He | 0 | $7 \times 10^{16}$ | 6,250 | Poueyo-Verwaerde |
| | | | 1.0 | $3 \times 10^{16}$ | 5,200 | et al. (1993) |
| | 15 | | 0 | $8.2 \times 10^{16}$ | 6,500 | |
| | | | 2.0 | $4.4 \times 10^{16}$ | 6,250 | |
| | | | 5.0 | $3 \times 10^{16}$ | 5,500 | |
| Steel | 10 $-$ | Ar | 0 | $1.1 \times 10^{17}$ | 12,900 | Seidel, Beersick, and Beyer (1994) |
| Al-Mg alloy | $10^6$ W/cm$^2$ (Nd:YAG) | Air | | $1.9 \times 10^{13}$ | 3,280 | Matsunawa et al. (1995) |
| Al | $4 \times 10^6$ W/cm$^2$ | $N_2$ | | $6.5 \times 10^{16}$ | 17,400 | Sokolowski, |
| | | He | | $4.6 \times 10^{15}$ | 16,100 | Herziger, and Beyer (1989) |
| | | $O_2$ | | $6.2 \times 10^{15}$ | 17,000 | |
| Al 2024 | $5 \times 10^7$ W/cm$^2$ (600 nm) | Vac | 0.5 | $3.3 \times 10^{17}$ | 81,00 | Knudtson, Green, and Sutton (1987) |
| | $1.3 \times 10^7$ W/cm$^2$ | | | $3.5 \times 10^{17}$ | 7,660 | |
| | $5 \times 10^7$ W/cm$^2$ | | 5 | $8 \times 10^{16}$ | 7,000 | |
| AlMgSil F28 | $1 \times 4$ av (pulsed $CO_2$) | Ar He | 1.2 | $3 \times 10^{17}$ $1 \times 10^{17}$ | | Michaelis et al. (1990) |

h is the height above the surface of the target.

and $T_e$ decrease with distance away from the laser focus but can still be large several millimeters off the surface, particularly at high incident laser intensity. Under certain high-power conditions, however, a stationary absorbing plasma can be formed in which $n_e$ and $T_e$ are constant over a distance of several centimeters in front of the laser focus (Seidel Beersick, and Beyer 1994). Under these conditions, an appreciable fraction (~60%) of incident laser power is absorbed in the plasma, leading to strong plasma heating and the maintenance of an extended region of high excitation at large distances away from the laser focus. The conditions in this plasma are similar to those in self-sustained Ar plasmas heated by $CO_2$ laser radiation (Mazumder, Rockstroh, and Krier 1987, Zerkle and Krier 1994).

A reduction in the ionization potential of the ambient gas by, for example, changing from He to Ar has the effect of increasing both $n_e$ and $T_e$, in the laser plasma, although the overall effect is not as substantial as one might expect (Table 8.3). A large reduction in these quantities is, however, observed in welding with Nd:YAG laser radiation. Indeed, the level of ionization and low electron temperature reported by Matsunawa et al (1995) in the plasma over Al-Mg alloys during welding with Nd:YAG radiation is more compatible with a high-temperature thermally excited gas than with a plasma. This effect arises from the wavelength dependence of the inverse Bremsstrahlung coefficient (equation 8.10), which ensures that direct heating of the plasma by 1.06-μm radiation is suppressed relative to that occurring at a wavelength of 10.6 μm.

Additional plasma heating terms such, as photoionization and multiphoton absorption can become important as the laser wavelength decreases. For example, Knudtson, Green, and Sutton (1987) found $n_e \sim 3 \times 10^{17}$ cm$^{-3}$ and $T_e \sim 8000°$K in an Al plasma produced by focussing 600-nm pulses from a dye laser on an Al target in vacuum. Under these conditions, inverse Bremsstrahlung is negligible, but Al atoms can be photoionized in a three-photon process. Because the cross section for this process will increase as $I^3$, where $I$ is laser intensity, three-photon ionization will be a dominant heating mechanism at high laser intensities. Knudtson, Green, and Sutton (1987) report on strong continuum emission from the plasma under these conditions ($I \sim 5 \times 10^7$ W/cm$^{-2}$) and attribute this to photons emitted in the recombination $Al^+ + e \rightarrow Al^* + hv$, where $Al^*$ is an excited state of the neutral Al atom.


## 8.3   PROPERTIES OF LASER WELDING PLASMAS

Plasmas are formed during welding of metals with $CO_2$ laser radiation when the laser intensity at the focus exceeds ~$10^6$ W/cm$^2$. As we have seen at such intensities, the plasma has a low degree of ionization and is attached to the surface of the workpiece, where it enhances coupling of incident laser radiation. The plasma is seeded by atoms evaporating from the workpiece, which generally have a lower ionization potential than that of the shield gas. The formation of the plasma can be inhibited to some extent by providing an excess of an inert gas such as He, but it mainly arises through ionization of the metal atoms.

At somewhat higher laser intensities or at slower weld speeds at which vaporization is enhanced, the density of metal atoms is sufficient to promote gas breakdown
with gas heating through inverse Bremsstrahlung. The plasma then becomes absorbing at the laser wavelength so that the surface of the workpiece is partially shielded.
Plasma heating leads to rapid energy deposition and the detachment of the plasma
from the surface of the workpiece. The plasma expands away from the workpiece
at $10^5$–$10^6$ cm/sec (Figure 8.7). The transient nature of this process introduces a
time-dependent shielding term into the laser intensity reaching the workpiece (Figure
8.8). This process occurs even with a CW laser intensity and introduces an instability
into the welding process. In reality, the laser intensity is rarely an entirely CW
signal because a variety of source instabilities yield output power fluctuations. These
fluctuations can combine with plasma shielding to produce strong time-dependent
interactions between the laser beam and the workpiece. These changes in coupling
can be the source of varying penetration depth, porosity, and weld morphology.

Calculation of the inverse Bremsstrahlung absorption coefficient shows that the
attenuation of incident laser radiation by absorption in the plume is not a strong
effect in these unstable plasmas. Heating and expansion of the plasma do, however,
result in refractive index gradients within the plume (Ducharme et al. 1992,
Heidecker et al. 1988, Matsunawa 1990, Miyamoto and Maruo 1992, Essien,
Keicher, and Jellison 1995). These gradients produce changes in focal spot size
(Poueyo et al. 1992) and therefore will alter the coupling between the laser beam
and the workpiece. Essien, Keicher, and Jellison (1995) used Schlieren imaging
techniques to follow these changes and found that refraction is greatest at the edge
of the plume (i.e., at the interface between the shock front and the surrounding
medium). A plot of the angle of refraction of 10.6-μm laser radiation transverse to

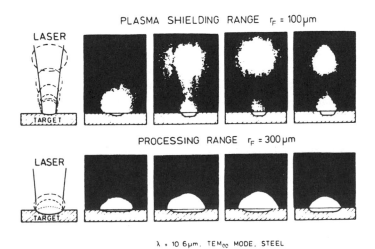

**Figure 8.7.** *Temporal development of laser plasma. (Top) high intensity; plasma detaches from
workpiece. (Bottom) low intensity; plasma is attached to workpiece. From Herziger, Kreutz, and
Wissenbach (1986).*

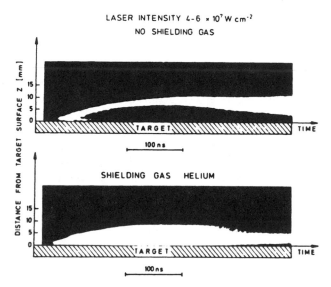

LASER INTENSITY 4-6 × 10⁷ W cm⁻²
NO SHIELDING GAS

**Figure 8.8.** *Streak photograph of the laser-induced plasma during welding of steel with $CO_2$ laser radiation. From Herziger, Kreutz, and Wissenbach (1986).*

the laser plume is shown in Figure 8.9 (Miyamoto and Maruo 1992) and shows that beam deflection is least on the axis of the plume. The effect of these gradients in refractive index is to introduce a diverging "lens" into the region of space over the workpiece. The focal length of this lens will be time dependent.

There is experimental evidence (Matsunawa 1990, Miyamoto et al. 1994) that ultrafine particles may be formed within the laser plume. At laser wavelengths of 1.06 and 10.6 μm, such particles will exhibit Rayleigh scattering, leading to strong attenuation under certain conditions (Hansen and Duley 1994). The scattering cross section scales as $\lambda^{-4}$, so this effect is much stronger at Nd:YAG wavelengths than for $CO_2$ laser radiation. It is uncertain as to the role that such particles play in shielding of the surface during laser welding. Table 8.4 summarizes the results of a detailed calculation of the attenuation of incident laser radiation by an aerosol of Fe particles when a range of particle sizes is present.

The overall attenuation of 10.6-μm laser radiation during welding of steel has been measured in a careful series of experiments by Miyamoto, Maruo, and Arata (1986). The intensity distribution in the transmitted beam during full penetration welding of thin plate (<1 mm) was measured through imaging (Figure 8.10). The beam is attenuated by a combination of Fresnel absorption at the surface of the keyhole and through absorption and scattering in the plume and keyhole plasmas. With a laser power of 1 kW, ~50% of power incident on a 0.12-mm-thick steel sheet was transmitted through the workpiece at welding speeds of 15–30 m/min. This was found to be consistent with a Fresnel absorption coefficient of ~50% and a plasma absorption coefficient, $\alpha$, of ~0–2 cm⁻¹. If $\alpha$ is identified primarily with absorption in the keyhole plasma, then this measurement suggests that plasma absorption is at most a small

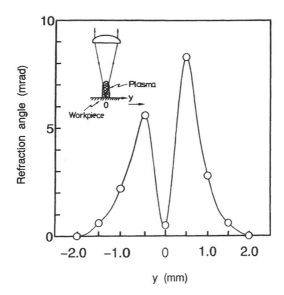

**Figure 8.9.** *Refraction angle $\theta$ for probe $CO_2$ laser beam passing transversely through the laser plasma created during $CO_2$ laser (2 kW) welding of steel at 50 cm/min in a 30 ℓ/min flow of Ar. From Miyamoto and Maruo (1992).*

effect. Miyamoto, Maruo, and Arata (1986) found, however, that plasma absorption becomes dominant when welding heavier gauge steel at a slow speed. In this case, attenuation of incident laser radiation by Fresnel and plasma absorption terms becomes comparable with $\alpha \sim 1–1.5$ cm$^{-1}$. The value of $\alpha$ measured in this way was found to decrease with a reduction in ambient pressure to a value of $\sim 0.5$ cm$^{-1}$ at $10^{-3}–10^{-4}$ Torr. Verwaerde et al. (1995) studied the effect of a reduction in ambient gas pressure on plasma properties and found that the laser-induced plasma virtually

**TABLE 8.4. Ratio of Transmitted Laser Intensity, *I*, to Incident Intensity, $I_0$, for a Path Length of $10^{-2}$ m in an Fe Aerosol in Which Particle Concentrations Follow an $r^{-3}$ Distribution, in Which r is Particle Radius**

| | | $I/I_0$ | |
|---|---|---|---|
| $\lambda$ ($\mu$m) | Size Range | $M/\rho = 10^{-4}$ | $M\rho = 10^{-6}$ |
| 1.06 | Small | 0.01 | 0.96 |
| | Medium | 0.00 | 0.95 |
| | Large | 0.65 | 0.99 |
| 10.6 | Small | 0.97 | 1.00 |
| | Medium | 0.88 | 1.00 |
| | Large | 0.96 | 1.00 |

$M/\rho$ is the ratio of the mass density of Fe particles in the aerosol to that of the solid. Small, medium, and large refer to distributions with an average r of 55 nm, 550 nm, and 5.5 $\mu$m, respectively.
From Hansen and Duley (1994).

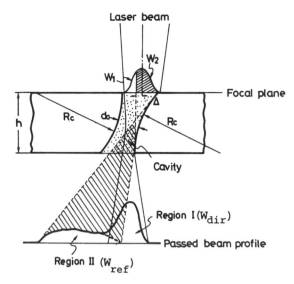

**Figure 8.10.** Geometry of incident and transmitted beam during full penetration welding of a thin steel sheet. From Miyamoto, Maruo, and Arata (1986).

disappears at pressures of ≤20 Torr. The electron density in the plasma was found to decrease by an order of magnitude from 760 to 20 Torr, whereas $T_e$ decreased from 6200 to 5500°K. Suppression of the plume by a reduction in ambient pressure has been shown to be effective in enhancing weld penetration (Arata 1987, Ishide et al. 1987, Verwaerde et al. 1995, Takano et al. 1990).

The profound effect that the shielding gas can have on weld properties under conditions where plasma formation occurs can be seen in Figure 8.11. At a laser power of 12 kW and with a welding speed of 1 m/min, the highest penetration was obtained with He as a shield gas, but this was accompanied by porosity. With $N_2$ gas, this penetration depth was a strong function of time but could reach ~3 mm. Penetration depth was limited to ~0.5 mm with Ar as a shielding gas. The sensitivity of penetration depth to shield gas composition was found to be directly related to the properties of the plasma formed (Table 8.5). Similar results were found with Nd:YAG radiation (Takano et al. 1990), although absorption of laser radiation in the plasma is less important at 1.06 $\mu$m, and detaching plasmas were observed for all three gases.

McCay et al. (1989) reported a comprehensive experimental study on the relation between plasma properties and weld characteristics for several Al alloys. Spectroscopic analysis was used to obtain the electron temperature $T_e$ in the plasma. With Ar as a shield gas, $T_e$ showed little dependence on flow rate, whereas a strong effect was observed with He. In the latter case, $T_e$ increased by ~50% as the flow rate varied from 0 to 13,800 cm³/min. The volatility and ionization potential of alloy components were found to have significant effects on $T_e$.

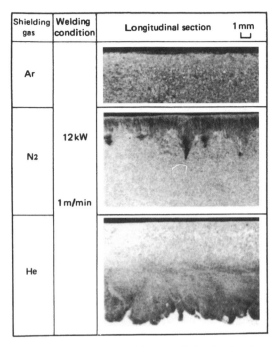

| Shielding gas | Welding condition | Longitudinal section    1 mm |
|---|---|---|
| Ar | | |
| N2 | 12 kW  1 m/min | |
| He | | |

**Figure 8.11.** *Longitudinal section in 304 stainless steel showing bead-on-plate weld at 1 mm using 12-kW $CO_2$ laser radiation and various shield gases. From Ishide et al. (1987).*

Under welding conditions, the formation of a plasma may be controlled to a certain extent through a secondary gas flow used to displace the plasma from its position over the workpiece (Miyamoto et al. 1984, Chennat and Albright 1984). Sun et al. (1993) showed how control over the flow of a transverse gas flow can result in a dramatic reduction in weld porosity and enhance weld penetration. Transverse flow rates of 10–20 $\ell$/min were found to be effective.

The angle that the secondary gas flow makes with respect to the incident laser beam has been shown to have a significant effect on weld cross-sectional area and penetra-

**TABLE 8.5. Summary of Relation Among Shield Gas Composition, Plasma Properties, and Weld Properties**

| Shield Gas | Plasma Properties | Weld Characteristics |
|---|---|---|
| He penetration | Continuous, attached to surface, low excitation | High porosity |
| Ar penetration | Floating plasma, 3–5 cm above surface | Poor porosity |
| $N_2$ penetration | Intermittent plasma, detaching on 3–5-msec time scale | Poor, strong spiking, low porosity |

Bead-on-plate weld in SS304 at 1 m min with 12 kW of $CO_2$ laser radiation.
Data from Ishide et al. (1987).

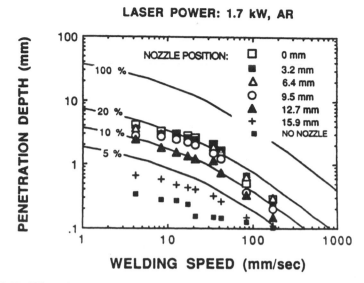

**Figure 8.12.** *Effect of plume suppression conditions on the penetration depth compared with penetration limit. The nozzle was oriented to provide an Ar gas flow transverse to the welding direction. From Chiang and Albright (1992).*

tion (Miyamoto, Maruo, and Arata 1984). The height of the nozzle above the workpiece has also been found to have a strong influence on weld penetration (Chiang and Albright 1992) and heat transfer efficiency; this can be seen in Figure 8.12.

Optimized conditions also can be achieved with a mixture of Ar and He (Seidel, Beersick, and Beyer 1994). Figure 8.13 shows the effect of a variation of the Ar/He

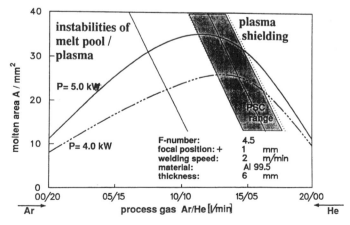

**Figure 8.13.** *Dependence of melt pool area on Ar/He ratio and gas flow rate in $CO_2$ laser welding of Al. From Seidel, Beersick, and Beyer (1994).*

ratio and flow rate on the area of the weld pool. An operating range at which real-time plasma shielding control (PSC) is effective through adjustment of laser intensity also is shown in this figure. These experiments show that optimization of shield gas composition, together with dynamic control based on sensing of plasma characteristics such as electron temperature, can yield high-quality welds with little porosity and with uniform penetration.

## 8.4 PLASMA SIGNALS FOR DIAGNOSTICS AND CONTROL

Because plasmas are created only when vaporization occurs, their presence during laser welding may be used to infer information about the welding condition. On the other hand, as plasmas signal vaporization and, specifically, the interaction between vaporized material and an incident laser beam, the information obtained through analysis of plasma radiation is only indirectly related to the laser material interaction that results in welding. The use of plasma diagnostics in monitoring laser welding then relies on the availability of a model that relates the observed variable to some aspect of the weld condition.

Monitoring of spectral line emission from a plasma produced during laser welding is a common method of diagnostics and yields information on the presence of a plasma. The intensity of such a spectral line, when monitored as a function of time, can be used to define a "set point" for the laser welding condition and for the detection of weld defects that are accompanied by a change in the intensity of plasma emission. Fast Fourier transform (FFT) techniques applied to an analysis of the temporal variation of spectral line intensity can be useful in identifying frequency components that are characteristic of specific weld faults. Time-resolved frequency spectra also can be used to sample keyhole dynamics.

A comparison of spectral line intensities provides information on plasma temperature, whereas spectral line width can be used to infer plasma electron density. Both of these parameters are important diagnostics of the plasma itself but may be difficult to relate to weld properties. They also are sensitive to the presence of impurities, particularly elements such as the alkali metals or the alkaline earths with low ionization potentials.

Simultaneous measurement of spectral line emission from the plasma together with infrared (IR) radiation from the weld pool itself provides a way to relate changes in plasma characteristics to the size and temperature of the weld pool. The reliability of this technique may be compromised by the fact that IR emission from the plasma also is present, making it difficult to separate IR signals emitted from the plasma from those arising at the surface of the weld pool. In addition, the thermal time constant for temperature changes in the weld pool can be much longer than plasma fluctuation time scales.

Prediction and monitoring of penetration depth during laser welding are important considerations in quality control and can be related to plasma characteristics. When feasible, the full penetration state can be monitored via plasma emission at the exit side of the weld. In many instances, this is impractical, however, and observations must be carried out from the direction of the laser beam. Imaging geometrics that involve looking down the keyhole have been shown to be useful in determining the position and temperature of the laser plasma within the keyhole and relating it to penetration depth. Direct measurement of the conductivity of the laser plasma using separate probes or by reference to the potential of the welding nozzle also are promising techniques.

The spatial extent and brightness of the plasma generated during laser welding can be observed via direct imaging with a CCD camera. The low framing rate (30 Hz) of standard cameras makes such monitoring of limited value, but higher framing rates combined with image analysis algorithms can yield useful diagnostic information. Morgan et al. (1996) imaged the laser focus during welding with Nd:YAG laser radiation by using a high-speed camera at a framing rate of up to 12 kHz and compared these images with time-dependent optical signals detected through the core of a fiber-optic delivery system. Correlation of these signals with keyhole properties, including the presence of defects, were investigated by Haran et al. (1996).

As it fluctuates, the laser plasma interacts with the surrounding gas and acts as a medium for transferring energy from the keyhole into the ambient medium. This interaction results in the generation of acoustic emission with temporal characteristics that are similar to those of the plasma. Monitoring of acoustic emission and filtering or analysis, via FFT techniques, of the frequency components of this emission also has been shown to contain information related to weld characteristics. A summary of this and other plasma diagnostic techniques used for monitoring and control of laser welding is given in Table 8.6.

A connection between the dynamic evolution of the plasma above the workpiece and spectral emission from the plasma has been well documented (Gatzweiler et al 1989, Beyer et al. 1987). Streak camera images show plasma fluctuations on time scales of $\leq 10^{-4}$ seconds but also show features that evolve on a much slower time scale ($\geq 1$ msec). Both of these occur as material is ejected from the keyhole and is heated through interaction with the incident laser beam. Due to the nonlinear nature of the laser/material interaction, the ejection of matter is an unstable process, with the ejection rate subject to a variety of instabilities. These instabilities include fluctuations in the position and shape of the keyhole, fluctuations in laser output, and interruption of the beam by attenuation and scattering in the plume. The presence of a shielding gas flow also will introduce hydrodynamic instabilities.

The random nature of many of these variables introduces considerable uncertainty into possible correlations between plasma parameters such as electron temperature and emission intensity and "good welding conditions." For example, fluctuations in plasma density change the coupling of laser radiation into the workpiece and may cause the keyhole to collapse. This can trap gases and may lead to weld porosity as well as to an uneven penetration depth. Figure 8.14 shows a comparison between the time dependence of plasma emission and laser power in the welding of an Al alloy (Olsen et al. 1992) and compares the FFT spectra of the two signals.

With a fast detection system, fluctuations in the coupling of laser radiation to the workpiece can be minimized. This is the basis for the PSC system developed

**TABLE 8.6. Representative Plasma Diagnostic Techniques Used for Monitoring and Control of Laser Welding**

| Diagnostic | Parameter | Reference |
|---|---|---|
| Optical emission | Intensity UV/IR | Chen et al. (1991, 1993) |
| | | Beyer et al. (1991) |
| | | Heyn, Decker, and Wohlfahrt (1994) |
| | | Maiwa, Miyamoto, and Mori (1995) |
| | | Miyamoto and Mori (1995) |
| | | Haran et al. (1996) |
| Spectral line | Plasma temperature | Miyamoto and Maruo (1992) |
| | | Miyamoto et al. (1993) |
| | | Sokolowski, Herziger, and Beyer (1988) |
| | | Seidel, Beersick, and Beyer (1994) |
| Optical emission | Time dependence | Gatzweiler et al. (1989) |
| | | Sokolowski, Herziger, and Beyer (1988) |
| | | Olsen et al. (1992) |
| | | Zimmerman (1993) |
| | | Miyamoto and Mori (1995) |
| | | Haran et al. (1996) |
| | | Morgan et al. (1996) |
| Acoustic emission | Time dependence | Shi et al. (1992) |
| | | Mao, Kinsman, and Duley (1993) |
| | | Duley and Mao (1994) |
| | | Gu and Duley (1996) |
| Streak camera | Time/spatial dependence | Gatzweiler et al. (1989) |
| | | Beyer et al. (1994) |
| Optical emission | Exit plasma optical transients, feedback control | Maischer et al. (1993) |
| | | Otto, Deinzer, and Geiger (1994) |
| | | Seidel, Beersick, and Beyer (1994) |
| Scattering/transmission | Particles, electron density | Matsunawa (1990) |
| Imaging | Morphology | Miyamoto et al. (1993) |
| | | Kinsman and Duley (1993) |
| Plasma conductivity | Electron density | Li, Steen, and Modern (1990) |
| | | Shi et al. (1992) |

by Seidel, Beersick, and Beyer (1994), Seidel et al. (1993), and Otto, Deinzer, and Greiger (1994). With PSC, the intensity of one or more spectral lines emitted by the plasma is monitored and used as a diagnostic of plasma conditions (Figure 8.15). When a threshold emission is reached, a correction signal is generated that acts to interrupt the laser intensity. The plasma then dissipates, and the laser intensity is allowed to increase again. This technique has been shown to result in more uniform weld penetration, as well as a reduction in the consumption of shield gases (Seidel et al. 1993, Seidel, Beersick, and Beyer, 1994). A study of the length and frequency of such plasma interrupts (Zimmerman 1993) has shown how pore formation and weld penetration depth can be related to plasma properties.

Detection of the degree of penetration during laser welding is an important parameter in quality control. This information can be obtained by looking down the laser

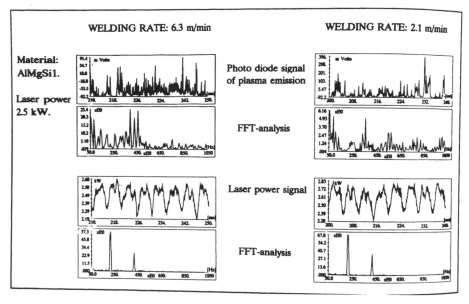

**Figure 8.14.** *Comparison of time dependence of plasma optical emission and laser power. FFT spectra also are shown. From Olsen et al. (1992).*

keyhole using a scraper mirror in the beam delivery system (Maischner et al. 1991, Beyer et al. 1991) or by using photodiodes mounted near the focussing mirror (Heyn, Decker, and Wohlfahrt 1994). When complete penetration is required, detectors may be mounted under the workpiece to sense emission from the keyhole as it exits the weld (Maischner et al. 1991). Miyamoto and Mori (1995) have shown that a comparison of optical emission from the keyhole plasma observed at oblique and near-

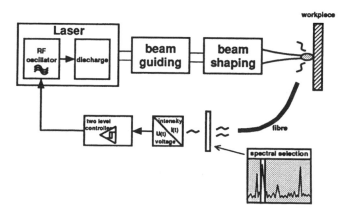

**Figure 8.15.** *Schematic of PSC system. From Seidel, Beersick, and Beyer (1994).*

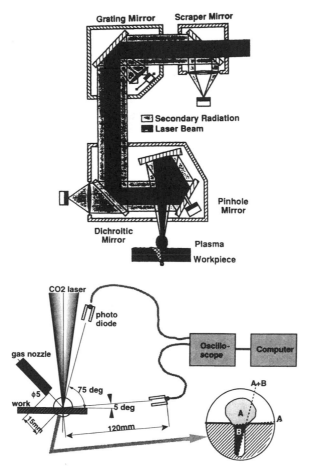

**Figure 8.16.** Weld penetration monitoring systems. (Top) from Beyer et al. (1991). (Bottom) from Miyamoto and Mori (1995).

normal incidence can be used to monitor penetration depth. Detector geometrics used in these monitoring systems are shown schematically in Figure 8.16. Kluft, Boerger, and Schwartz (1996) discussed the use of ultraviolet plasma detection in detection of penetration depth during $CO_2$ laser welding of automotive sheet steel.

The basic measurement in all of these systems is that of plasma emission intensity over one or more spectral regions. Changes in integrated intensity averaged over time can be used to infer welding conditions. The behavior of the keyhole plasma under partial and full penetration conditions is shown in Figure 8.17 (Miyamoto and Mori 1995). With the keyhole closed, the keyhole plasma is observed to be localized near the entrance and oscillates about this position. As a result, emission from the plasma above the keyhole and that inside it are out of phase, and the plasma is not free to expand toward the closed end of the keyhole.

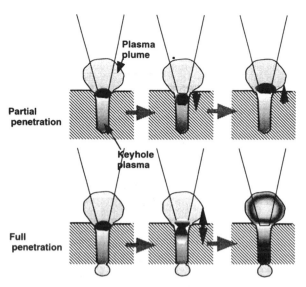

**Figure 8.17.** *Schematic of behavior of keyhole and plume plasmas under partial and full penetration conditions. From Miyamoto and Mori (1995).*

Under complete penetration conditions, the plasma can expand in both directions, so plasma emission detected above and inside the keyhole will be correlated. This behavior will be apparent through a variation in FFT spectra of signals obtained under fully penetrated and nonpenetrated welding conditions, but it may not be as obvious when comparing emission intensity.

The intensity of plasma emission as recorded by looking down the keyhole (Figure 8.16, top) or through imaging of the plasma at an angle relative to the laser beam (Figure 8.16, bottom) is most useful in detecting weld fault conditions because these commonly involve significant changes in total plasma emission. Feedback of light through the delivery fiber in welding with Nd:YAG radiation has been shown to be effective in monitoring welding conditions (Haran et al. 1996). Analysis of the frequency spectrum of the transients in these signals can provide additional information but adds an extra level of complication to the monitoring system.

Table 8.7 summarizes the response of experimental plasma emission detection systems to common defects and faults in laser welding. Although the effect of a fault or defect can produce a noticeable and reproducible change in the intensity of plasma emission, detailed behavior often is complex. This complexity can be seen, for example, in Figure 8.18, which shows a record of plasma emission detected with a photodiode sensor looking down the keyhole (P1) and compares this emission with that from a similar photodiode looking only at the plasma above the surface (P4). The difference P1–P4 increases in a region of underfilling of the butt weld and has been found to be linearly related to the degree of underfill (Miyamoto et al. 1993).

**TABLE 8.7. Detector Response for Various Weld Defects and Faults**

| Error | Detector Response | Reference |
|---|---|---|
| Gaps/displacement | Reduction in signal | Beyer et al. (1991) |
| Keyhole failure | UV, IR decrease | Chen et al. (1993) |
| Surface defect | UV, IR decrease | Chen et al. (1993) |
| Humping/penetration | UV, IR oscillate | Chen et al. (1993) |
| Edge misalignment | Enhanced near IR | Heyn, Decker, and Wohlfahrt (1994) |
| Pits | Difference between keyhole and plume plasma signals decreases | Miyamoto et al. (1993) |
| Underfill | Difference between keyhole and plume plasma signals increases | Miyamoto et al. (1993) |

UV, ultraviolet. IR, infrared.

The use of an FFT to obtain the spectrum of frequency components associated with "normal" welding may make it possible to selectively monitor specific weld fault conditions. Before this can be accomplished, the normal spectrum corresponding to optimized or desirable welding conditions must be characterized. Figure 8.19 shows FFT spectra of optical and acoustic emission obtained during $CO_2$ laser welding of mild steel at various weld speeds. The spectrum recorded at a speed of 3.8 cm/sec shows a dominant frequency component at ~4.5 kHz, whereas that at other speeds contains a wide range of components. Examination of weld cross sections obtained under these conditions has shown that minimum porosity and a small heat-affected zone are obtained when the spectrum shows only the 4.5-kHz peak. This may indicate optimized coupling of laser radiation into the keyhole and maintenance of a stable keyhole geometry (Gu and Duley 1996a).

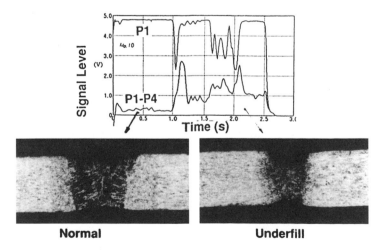

**Normal**          **Underfill**

*Figure 8.18.* *Record of plasma emission. P1, from keyhole versus time. The increase in P1 relative to that observed from the surface plasma P4 is related to the degree of underfill. From Miyamoto et al. (1993).*

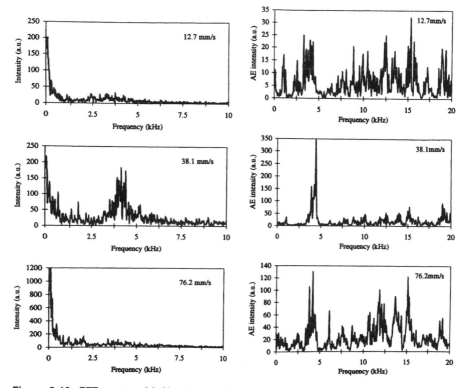

**Figure 8.19.** FFT spectra of (left) optical and (right) acoustic emission from plasma during welding of 1.14-mm-thick mild steel at an incident $CO_2$ laser intensity of $1.06 \times 10^6$ $W/cm^2$. From Gu and Duley (1996a).

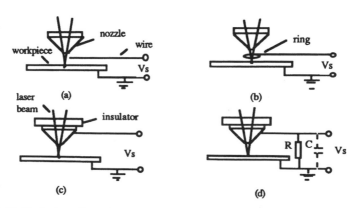

**Figure 8.20.** Typical configurations for electric probe measurement of plasma characteristics during laser welding. From Li et al. (1990).

A change in FFT spectrum also can be used to monitor focal position or a deviation from a preferred focus (Maischner et al. 1994) with an increase in randomness in the frequency spectrum accompanying detuning from optical focussing conditions.

A novel plasma diagnostic technique based on electrical probe measurements of the plasma conductivity was described by Li et al. (1990), Li and Steen (1992), Shi et al. (1992), and Steen (1991). A variety of electrical configurations are possible (Figure 8.20) and it has been shown (Li et al. 1990) that the signal generated can be used to identify various types of weld fault or weld misalignment errors. Sensitivity varies directly with applied voltage, and the response time in typical applications was found to be ~25 μsec. The incorporation of a sensor of this type into an expert system for welding of cans was reported by Shi et al. (1992).

## 8.5 ACOUSTIC EMISSION

A noticeable characteristic of laser welding is the acoustic emission that accompanies this process, particularly when a plasma is present. The range of frequencies emitted includes those in the audible range (20 Hz–20 kHz), so an experienced operator may monitor optimal welding conditions by listening to the tones emitted. Essentially, this process consists of Fourier transformation of the acoustic wave intensity, followed by the application of an algorithm that compares the amplitude within a range of detected frequencies with an expert database. Simulation of this procedure using an acoustic detector (other than the ear) offers the possibility of process diagnostics and control.

A number of physical processes contribute to the emission of acoustic energy during laser welding (Table 8.8). Because of strong coupling to the ambient gas, keyhole oscillations, vaporization, and plasma heating/expansion are the dominant contributors to airborne acoustic signals. These signals also may be detected via transducers attached to the workpiece, but in practice, contact sensors are used only in the detection of acoustic emission accompanying solidification and cracking.

The spectrum of acoustic emission during laser welding can extend to megahertz frequencies, but useful information can be obtained from spectra recorded over a much smaller range ($\leq 100$ kHz). This is particularly true under keyhole welding conditions when vaporization and plasma formation are closely related to instabilities

**TABLE 8.8. Sources of Acoustic Emission During Laser Welding**

| Process | Characteristics |
|---|---|
| Thermal expansion | Stress wave in workpiece |
| Solidification | Postweld |
| Vaporization | Pressure fluctuations in ambient gas |
| Keyhole oscillation | Couples mass motion in melt to gas |
| Plasma | High-pressure transients due to plasma heating and expansion |
| Particle ejection | Couples surface instability to gas |
| Gas flow | From shielding gas |

and mass motion in the melt surrounding the keyhole. The frequency of oscillations characterizing these processes are typically $\leq 10$ kHz, so acoustic spectra in the range of 100 Hz–20 kHz will be indicative of keyhole stability and welding conditions.

The driving term for acoustic emission is the excess pressure in the keyhole

$$\Delta p = p - p_0, \tag{8.31}$$

where $p$ is the keyhole pressure, and $p_0$ is ambient pressure. $p$ is associated with the vapor pressure of metal within the keyhole and is related to the mass evaporation rate $dm/dt$ (kg/m$^2$/sec) through

$$\frac{dm}{dt} = p(T)\left[\frac{\overline{m}}{2\pi kT}\right]^{1/2}, \tag{8.32}$$

where $\overline{m}$ (in kilograms) is the average mass of a vaporizing atom, and $p(T)$ is given by the Clausius-Clapeyron equation:

$$p(T) = p(T_B)\exp\left[\frac{L_v}{Nk}\left(\frac{1}{T_B} - \frac{1}{T}\right)\right], \tag{8.33}$$

where $T_B$ is the normal boiling temperature (°K), $L_v$ is the latent heat of vaporization (J/m$^3$°K), N is the atomic density in the solid (atoms/m$^3$), and $k$ is the Boltzmann constant (J/°K). The acoustic signal V(t) detected will be proportional to $dp/dt$. Then, with $C_1$ as a constant:

$$V(t) = C_1 \frac{dp}{dt}, \tag{8.34}$$

which becomes, using equation 8.32:

$$V(t) = C_1\left[\frac{p}{2T}\frac{dT}{dt} + \frac{p}{(dm/dt)}\frac{d}{dt}\left(\frac{dm}{dt}\right)\right]. \tag{8.35}$$

If the mass vaporization rate dm/dt arises through an oscillation of the keyhole at frequency, $f_i$ then

$$V(f_i,t) = pC_1\left[\frac{1}{2T}\frac{dT}{dt} + f_i\right] \tag{8.36}$$

where

$$\frac{dT}{dt} \simeq \frac{\Delta T_i}{\tau_i}, \tag{8.37}$$

and $\Delta T_i$ is the temperature excursion during the perturbation of the keyhole boundary for the $i$th keyhole frequency and $\tau_i$ is the damping time. Then:

$$V(f_i) \simeq pC_1\left[\frac{1}{2\tau_i}\frac{\Delta T_i}{T} + f_i\right] \qquad (8.38)$$

Because $\tau_i^{-1} < f_i$ and $\Delta T_i$ is expected to be comparable to $T$, the first term in the bracket will dominate except at high $f_i$, where $\tau_i^{-1} \simeq f_i$ and the two terms will become comparable. In either case, frequency components in the mass motion of the liquid surrounding the keyhole appear in the spectrum of the acoustic signal.

FFT processing of the acoustic emission during CW laser welding reveals a complex spectrum (Figure 8.21), with a variety of overlapping broad and sharp

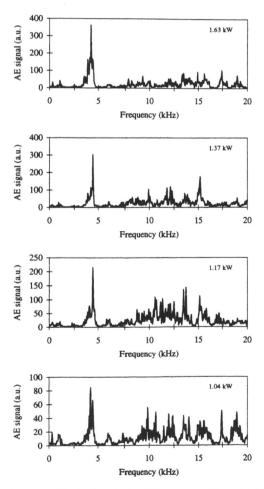

**Figure 8.21.** Acoustic energy distribution during $CO_2$ laser welding of 1.14-mm-thick mild steel at various laser powers for bead-on-plate weld at 3.8 cm/sec. From Gu (1995).

spectral features (Gu and Duley 1994). In general, the relative intensity of components in these FFT spectra are sensitive to a variety of parameters, including such factors as laser power, beam focussing, welding speed, surface preparation, metal composition, and shielding gas flow. Acoustic spectra illustrating some of this variability are shown in Figures 8.22–8.25. For convenience and to assist in the comparison of spectra obtained under different conditions, frequency components have been binned in 1-kHz intervals and normalized such that the total integrated spectral emission is unity. These show, for example, how the overall acoustic emission spectrum becomes less complex at certain welding speeds (Figure 8.24). Such changes in spectral distribution can be used in the development of process diagnostic and control procedures.

In $CO_2$ laser welding of Al 1100 (Figure 8.25), the spectrum of acoustic emission changes dramatically at incident laser powers of <1100 W. This is the threshold for welding under these conditions.

The use of acoustic emission as a diagnostic technique in laser welding was investigated as early as 1976 (Saifi and Vahaviolos 1976) in an application involving pulsed Nd:YAG laser welding of Cu wires to terminal posts. With a piezoelectric transducer in contact with the part, the amplitude of acoustic emission was measured and compared with a standard to determine the acceptability of individual compo-

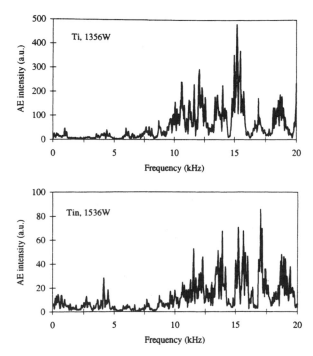

**Figure 8.22.** Acoustic emission spectra obtained during $CO_2$ laser welding of Ti and Sn for bead-on-plate at 3.8 cm/sec. From Gu (1995).

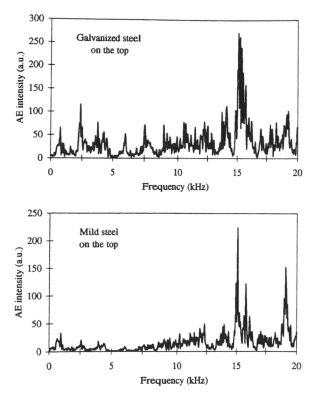

**Figure 8.23.** *Acoustic emission spectra obtained during $CO_2$ laser welding of steel. Lap weld, 1-mm-thick sheets, no gap, 3.8 cm/sec. From Gu (1995).*

nents. This pioneering work anticipated several of the techniques that are under evaluation for monitoring of weld quality (Habenicht, Stark, and Deimann 1991, Hanting and Aiquing 1992, Willmott, Hibbard, and Steen 1988, Schou, Semak, and McCay 1994, Farson et al. 1994). However, a measurement that relies on contact with the workpiece is inconvenient. Later studies have shown that similar process information can be obtained with noncontact acoustic sensors (Jon 1985, Hamann, Rosen, and Lassiger 1989, Steen and Weerasinghe 1986). A comprehensive study of the use of acoustic emission in quality assurance of spot welds (Habenicht, Stark, and Deimann 1991) has shown how FFT processing of acoustic signals yields much more information than a simple measurement of global acoustic emission intensity. They found, for example, that specific spectral signatures can be identified that correspond to the formation of a melt pool, generation of spatter, and emission due to postweld cracking. However, they also found that there was no correlation between acoustic emission components and the subsequent strength of the welded joint.

The relation among acoustic emission, penetration depth, and plasma emission was investigated by Willmott, Hibbard, and Steen (1988) and Schou, Semak, and McCay (1994). The experimental configuration used by Willmott, Hibbard, and

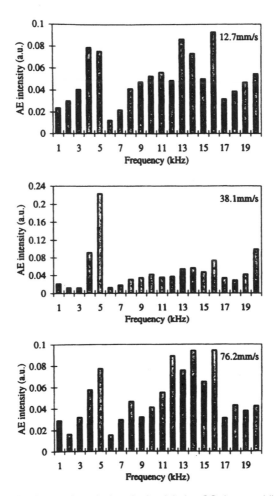

**Figure 8.24.** Normalized acoustic emission obtained during $CO_2$ laser welding of 1.14-mm-thick mild steel at 1.65 kW and various welding speeds. From Gu (1995).

Steen (1988) included an ultrasonic (1 MHz) sensor in contact with the final mirror in the beam delivery system plus an optical sensor mounted near the gas nozzle. The relative outputs from these detectors are plotted schematically in Figure 8.26 and show that the acoustic signal first rises before initiation of the keyhole. With a keyhole present, the overall acoustic emission drops while optical emission from the plasma is enhanced. Farson, Ali, and Sang (1997) and Farson, Sang, and Ali (1997) found a possible correlation between the derivative of the optical emission signal from the plasma and the amplitude of the acoustic signal.

Schou, Semak, and McCay (1994) observed a more complex behavior when laser pulses were used for spot welding. Three distinct stages were noted. The first was a transient acoustic emission that occurred promptly at the start of the laser pulse

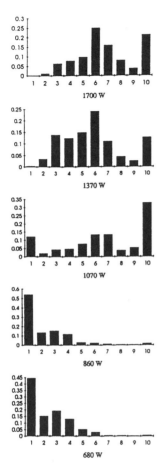

**Figure 8.25.** *Distribution of acoustic spectral components with laser power during welding of Al 1100 with CO₂ laser radiation at scan speed of 25.4 mm/sec. From Duley and Mao (1994).*

and could be identified with the vaporization of surface impurities and irregularities. This was followed at a later time by a low-intensity acoustic emission characteristic of the vibrations of the weld pool. Finally, a strong transient was observed that coincided with the onset of plasma ignition and the maintenance of a fully penetrated keyhole. FFT spectra of the acoustic emission obtained under nonpenetrated and fully penetrated conditions showed that complete penetration was characterized by a strong peak near 4 kHz. This peak was not present under nonpenetrated spot welding conditions, where the spectrum was more randomized. This behavior is similar to that shown in Figure 8.24 for seam welds and suggests that full penetration occurs only when the keyhole geometry is optimized. The 4-kHz frequency component was unaffected by a change in cover gas from He to Ar, suggesting it represents a natural or eigenfrequency of the keyhole–weld pool boundary layer.

Discrimination between ''good'' and ''bad'' welds by analysis of acoustic signals can be accomplished only through signal processing followed by the application of an algorithm that infers the subset of correct conditions from the data set presented.

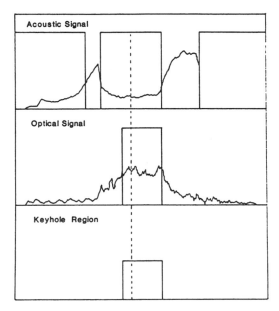

**Figure 8.26.** Time dependence of acoustic and optical emission signals related to presence of keyhole. From Willmott, Hibbard, and Steen (1988).

This approach was investigated by several groups (Shi et al. 1992, Farson, Fang, and Kern 1991, Farson et al. 1994, Gu and Duley 1996b). Spectral components in the region between 1 and 5 kHz seem to be particularly useful as diagnostic of weld penetration. The reliability of weld classification based on a comparison of acoustic energy within this band with that elsewhere in the spectrum can be excellent (Table

**TABLE 8.9. Classifier Performance for Weld Penetration Based on Acoustic Spectrum**

| Weld No. | Laser Power (kW) | Travel Speed (in/min) | Actual Penetration | Prediction Reliability (%) | Averaged Prediction Reliability (%) |
|---|---|---|---|---|---|
| 1 | 8 | 110 | Full | 91.3 | 98.3 |
| 2 | 8 | 110 | Full | 96.8 | 100 |
| 3 | 8 | 110 | Full | 100 | 100 |
| 4 | 8 | 110 | Full | 100 | 100 |
| 5 | 8 | 180 | Partial | 66.7 | 70.8 |
| 6 | 8 | 180 | Partial | 97.2 | 100 |
| 7 | 8 | 180 | Partial | 90.5 | 100 |
| 8 | 8 | 180 | Partial | 96.6 | 100 |
| 9 | 4 | 110 | Partial | 95.4 | 99.2 |

Based on Farson et al. (1994).

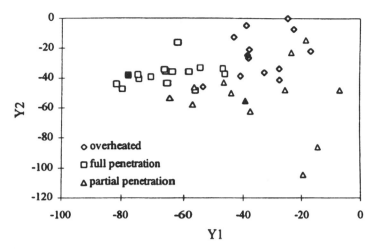

**Figure 8.27.** *Statistical qualification of laser welds in mild steel based on acoustic spectral components. The average solution for fully penetrated welds (filled squares) is well separated from that for overheated or partial penetration welds (filled diamond and filled triangle, respectively). From Gu and Duley (1996b).*

8.9). Figure 8.27 shows a comparison of solutions in discriminant space based on a comparison of measured acoustic spectral data with that of a training set. The filled symbols indicate average values for three weld classifications corresponding to overheated (i.e., large heat-affected zone), fully penetrated, and partially penetration welds in mild steel (Gu and Duley 1996b,c).

The monitoring of both optical emission from the keyhole and accompanying acoustic emission has been demonstrated to provide additional diagnostic capability, particularly with regard to the determination of penetration depth (Beersiek et al. 1997). Because optical emission can be linearly related to penetration depth and acoustic emission shows a minimum at a given penetration depth, a combination of these signals can provide a useful predictor of weld depth. An example of such an optical-acoustic discriminator in the welding of mild steel with $CO_2$ laser radiation is shown in Figure 8.28. Here, the penetration depth could be determined to an accuracy of $>10\%$. Acceptable welds were obtained only for the narrow region between 1.0 and 1.1 mm, where both acoustic and optical signals satisfied their individual criteria.

## 8.6   VISUAL MONITORING

The direct imaging of the weld pool and the surrounding area on the workpiece during laser welding was reported by several groups (Denney and Metzbower 1991, Schimon and Mazumder 1993, Kinsman and Duley 1993, Li, Steen, and Modem 1993, Griebsch et al. 1994a, 1994b, Semak et al. 1994a,b, 1995a, 1995b, Bagger et

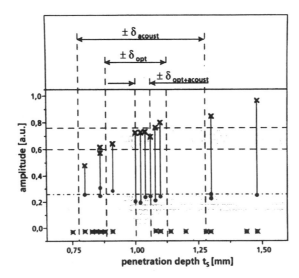

**Figure 8.28.** Comparison of acoustic and optical signals in $CO_2$ laser welding of mild steel. Both optical and acoustic signals must satisfy criteria for the weld penetration to be acceptable. From Beersiek et al. (1997).

al. 1992, Beersiek et al. 1997). The simplest approach is to use a standard CCD camera with no external illumination (Li, Steen, and Modem 1993, Kinsman and Duley 1993), but this yields only limited data because emission from the plasma plume often obscures the surface of the workpiece. To a certain extent, this effect can be minimized through the use of an optical filter tuned to a spectral region at which the plasma emission is small; however, this also reduces the amount of available light. The slow framing speed of standard CCD cameras (30 Hz) limits the overall ability of the system to follow rapid changes in weld pool morphology or the dynamics of mass motion.

The first problem can be overcome through illumination of the weld pool with CW laser radiation from, for example an Ar ion laser plus the use of a narrow band interference filter centered on the laser line (Semak et al. 1995a,b). Inspection of temporal changes in the laser material interaction and weld pool morphology requires high-speed imaging using CCD (Griebsch et al. 1994a, 1994b) or photographic (Semak et al. 1994b) techniques. Synchronization of a pulsed laser illuminator with a high-speed video system yields optimized imaging of surface phenomena (Denney and Metzbower 1991, Griebsch et al. 1994a, 1994b). A schematic of such a system is shown in Figure 8.29. In this system, the video camera sends a TTL signal to the trigger generator, which initiates the computer and triggers a pulse from the frequency doubled Q-switched Nd:YAG laser illuminator. Photodiode and fast laser power monitor signals also are digitized by an analog-digital board on the computer. The overall temporal resolution in imaging was 1 msec.

Griebsch et al. (1994a, 1994b) used this system to study droplet formation and

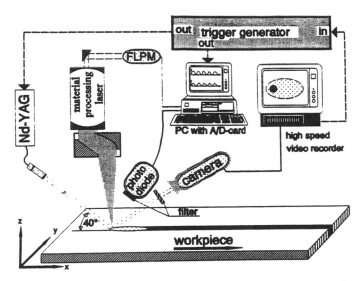

**Figure 8.29.** *Experimental system for fast imaging (1 kHz) of weld phenomena. From Griebsch et al. (1994a).*

ejection during $CO_2$ and Nd:YAG laser welding. Both sources were operated in a pulsed mode, with pulse durations of 3.5–3.7 msec (YAG) and 0.5–18 msec ($CO_2$). They found that droplets are generated when the interval between laser pulses is sufficiently large to allow solidification. Droplets also are formed when the peak power is too large. In general, droplet emission can be minimized when the weld pool volume is kept large. An example of the correlation between droplet ejection and melt pool morphology can be seen in Figure 8.30. It is apparent that weld quality is degraded when large droplets are formed. An analysis of droplet formation and ejection in relation to the dynamics and melt motion in the weld pool shows that this arises in accelerations of the weld pool driven by vaporization. A large weld pool volume, together with low laser pulse intensity, minimizes this effect and inhibits the ejection of droplets (Griebsch et al. 1994a, 1994b).

The oscillation of the weld pool surface also is apparent in these data and takes the form of surface waves, often of appreciable amplitude, that radiate from the keyhole. Semak et al. (1995a) studied these oscillations using Schlieren photography and laser beam reflection. Spot welds were produced with single millisecond-duration pulses from a $CO_2$ laser irradiating 304 stainless steel. The Schlieren system used a high speed ($10^4$ Hz) framing camera with an exposure time of 10 $\mu$sec. These observations detected surface oscillations at frequencies of ~200–300 Hz during the laser pulse, with amplitudes comparable to the width of the melt pool. Droplets were found to be ejected at velocities of ~1 m/sec under these conditions. This motion ceased 0.5–1 msec after the termination of the laser pulse and the amplitude of surface oscillation diminished. After the laser pulse, the frequency of surface waves

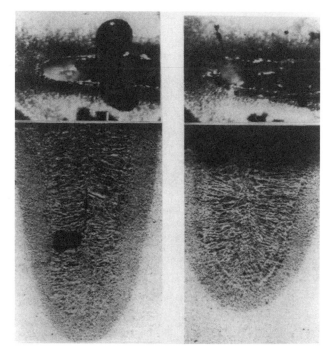

**Figure 8.30.** (Top) droplet ejection from melt pool and resulting weld structure (bottom) during pulsed $CO_2$ laser welding at high average power (left) and low average power (right). From Griebsch et al. (1994a).

increased dramatically into the 1.5-kHz range as the amplitude of these waves reduced. The origin of this high-frequency oscillation is uncertain.

A similar experimental configuration was used by Semak et al. (1995b) to image the top of the weld pool during CW $CO_2$ laser welding of 304 stainless steel. These measurements showed that the diameter of the entrance hole to the keyhole exceeds the laser spot size at slow welding speed but then is reduced to the beam diameter at higher speeds. The keyhole was also found to trail the laser spot at high weld speed. Under these conditions, only the front wall of the keyhole was found to be exposed to incident laser radiation.

High-speed video imaging of lap welds in mild steel and Zn-coated steel (Bagger et al. 1992) has demonstrated how characteristics of the total light emission can be related to weld properties (Figure 8.31). Nonpenetrating surface welds are accompanied by weak light emission localized near the laser spot on the surface of the top sheet. Coupling is enhanced at higher incident laser intensity, leading to vaporization and the appearance of a plume over the workpiece. This still is a nonpenetrating condition because no keyhole has been formed.

Unstable keyhole welding with fluctuations in penetration depth shows rapid variations in weld pool area and brightness of the surface plasma. Complete penetration is characterized by the appearance of a plasma on the backside of the lap weld, as well as a smaller weld bead width and area. The emission intensity under these conditions was found to be similar to that observed with vaporizing surface welds.

In a similar experiment, Schimon and Mazumder (1993) studied the relation between the dimensions of the surface melt pool and the interface bead width in lap welds. A good correlation between these two parameters was observed, suggesting visual imaging of the bead width may be a useful diagnostic of interface weld properties.

An experiment in which the keyhole cross section transverse to the optical axis is imaged with a fast CCD camera looking down the laser beam was reported by Beersiek et al. (1997). Such images show that the keyhole profile is correlated with welding speed and depth of penetration. This information can be used to estimate the keyhole shape and its location within the workpiece (Figure 8.32). The keyhole cross section is found to narrow considerably at certain depths during welding under complete penetration conditions, making this a useful technique for monitoring penetration depth during lap or T-joint welding.

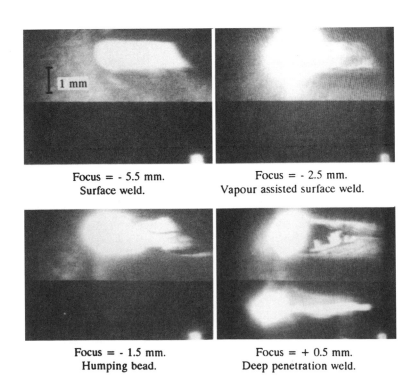

Focus = - 5.5 mm.
Surface weld.

Focus = - 2.5 mm.
Vapour assisted surface weld.

Focus = - 1.5 mm.
Humping bead.

Focus = + 0.5 mm.
Deep penetration weld.

**Figure 8.31.** *Video frames showing dependence of light emission on weld characteristics. (Top) from the front surface of the lap weld. (Bottom) from the back surface. From Bagger et al. (1992).*

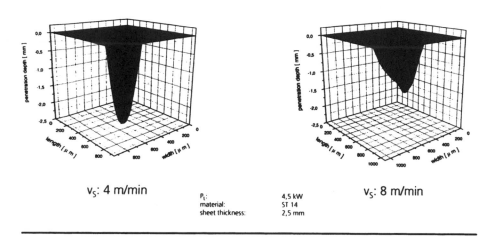

**Figure 8.32.** Shape of the keyhole during $CO_2$ laser welding of steel as determined from fast CCD imaging of keyhole cross section. From Beersiek et al. (1997).

# 9

## Control of Laser Welding

### 9.1  INTRODUCTION

A variety of signals are generated during laser welding, and these signals contain information about the welding process. They contain information about what is "right" about the process, as well as what is "wrong." Extraction and recognition of signal components that characterize "right" and "wrong" is the key to process monitoring and, subsequently, process control.

The definitions of "right" and "wrong" in relation to a characteristic of the laser welding operation depend in turn on what are acceptable operational limits and what are to be considered to be defects. The identification and definition of specific defects are fundamental to monitoring for process control because weld defects that are critical to one application may be unimportant in another.

The basic philosophy therefore is to attempt to answer the question, "For what do we control?" Although many weld defects often are correlated, the identification of specific critical defects acts to define monitoring strategy, set protocols for process feedback, and control for optimization of weld quality.

### 9.2  WELD DEFECTS

The weld defects that are important in a particular application depend on the service requirements of the weld. These in turn will depend on environment, joint configuration, postweld forming operations, heat treatment, and other parameters. Susceptibility to particular types of weld defect is, however, linked to material composition. Some general guidelines relating alloy type to common defects are summa-

**TABLE 9.1. Common Weld Defects vs. Alloy Type**

| Alloy | Porosity | Solidification Cracking | Cold Cracking | Corrosion Cracking |
|---|---|---|---|---|
| Low-carbon, low-alloy steels | | | X | |
| Medium-, high-carbon steels | X | X | X | X |
| High-alloy steels | | X | X | X |
| Aluminum alloys | X | X | | |
| Titanium alloys | | | X | |
| Nickel alloys | X | X | | |

rized in Table 9.1. This list is by no means comprehensive but does illustrate common problems that may arise during welding materials of a given composition.

Weld defects also commonly arise through faults in joint characteristics such as fit up and alignment, these defects include incomplete penetration, burn through, slumping, pitting, porosity, undercutting, and others. Specific fault conditions may arise with special combinations of joint configuration and materials. For example, lap welding of galvanized steels requires a gap between sheets to allow the escape of Zn vapor and prevent separation at the join during welding.

A summary of common fault conditions related to fit up and joint configuration is given in Table 9.2. The common butt weld configuration, which is of importance in tailor blanking, is susceptible to misalignment of the sheets either across the gap or perpendicular to it (Figure 9.1). This places constraints on the clamping geometry and may require tacking of the seam ahead of the weld location to limit seam separation in response to stresses induced by the welding operation.

In addition to intrinsic defects attributable to material properties and those that arise from problems associated with joint design, there are a number of fault conditions that can be assigned to the laser welding process itself. Several of these were identified by Steen (1991) and Dawes (1992) and are summarized in Table 9.3. A useful map of weld bead profile versus laser power and weld speed for thin sheets of 304 stainless and carbon steel was derived by Albright and Chiang (1988b) and

**TABLE 9.2. Fault Conditions in Laser Welding with Various Joint Configuration**

| Joint Configuration | Defect | Fault |
|---|---|---|
| Butt | Gap fit up* | Burn through, lack of weld, undercutting |
| | Surface fit up† | Little effect if <50% of smallest sheet thickness |
| | Burrs | Burn through, porosity |
| Lap | Gap between sheets | Slumping, loss of penetration |
| Lap edge/T edge | Same as for butt | Breakout to side |
| T butt/lap fillet | Laser beam/joint alignment | Lack of penetration in both components |

* See Figure 9.1a.

† Figure 9.1b.

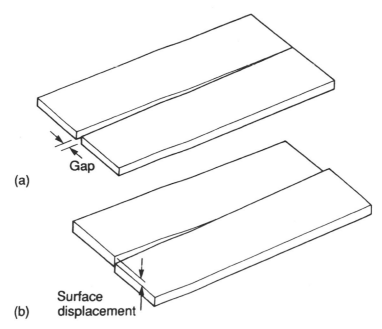

*Figure 9.1.* Fit up defects in butt welding: (a) gap mismatch. (b) surface mismatch. From Dawes (1992).

shows how process variables can be adjusted to operate in the range of an acceptable weld condition.

During laser welding, the fault conditions defined in Tables 9.2 and 9.3 are superimposed on the tendency for a particular alloy to exhibit specific intrinsic weld defects. Any protocols for identification and possible correction of a weld fault during processing must then include recognition algorithms for each of the defects considered to be critical to the weld being produced. This is a formidable task, perhaps best accomplished in a fuzzy logic/neural net environment.

**TABLE 9.3. Weld Defects and Fault Conditions Attributable to the Laser Welding Process**

| Condition | Source |
|---|---|
| Lack of penetration | Fluctuation in laser power, defocussing of beam, dirt on optics, focus moved off seam, inadequate shielding |
| Porosity, pits, blow holes | Gas shielding, composition, and flow rate; oxidation; contamination of metal; dross buildup; fluctuation in laser power; vaporization of coatings |
| Humping, undercut | Weld speed too high for power/thickness combination |
| Drop out, slumping | Laser power too high, speed too slow; excessive gas flow |
| Weld spatter | Laser power too high, oxidation, contamination, speed too low |

Two major questions that must be addressed in any monitoring and control system are: ''Which signal characteristics are diagnostic of particular weld defects and/or weld fault conditions?'' ''What modification of system parameters is required to respond to these signals in such a way as to return the system (i.e., the weld) to its desired state?''

A dominant consideration in answering these questions is the speed of response, namely the rate at which data can be collected, analyzed, and used to generate a correction signal. The response time of the system to the correction signal also is critical.

The behavior of ultraviolet (UV) infrared (IR) and acoustic signals in response to various defects and fault conditions was discussed in Chapter 8. The primary response is a change in level in the signals detected from one or more sensors. This may occur as a localized temporal spike, a slow drift in level, or a repetitive time varying oscillation. In general, a drift in signal level is characteristic of alignment defects such as problems with joint fit up, beam wander, and slow changes in focus. An increase in IR emission can signal a tendency to burn through in T or lap welds. Correlated drifts in UV and IR signals can be characteristic of beam focus or gas shielding problems.

Signal changes that accompany porosity and lack of penetration occur over rapid time scales and give rise to a range of frequency components when transformed from the time domain to frequency space. These components are generated as laser radiation interacts with the walls of the keyhole and the plasma produced in the vaporization of this material. They then reflect both the mechanical motion of the liquid and the mass flow within the keyhole. One expects different components of these signals to appear with specific time delays after an interactive event and, thus, time series of UV/IR and acoustic signals to be correlated. Some approximate time scales for signal changes that can be attributed to various physical processes are given in Table 9.4 (Schulz et al. 1996). Table 9.5 lists time scales associated with the presence of mechanical defects during laser welding.

**TABLE 9.4. Time Scales Related to Physical Processes during Laser Welding**

| Process | Time Scale (sec) |
|---|---|
| Plasma fluctuation | $<10^{-5}$ |
|    Optical/acoustic emission | |
|    Beam deflection | |
| Capilliary motion | $<10^{-4}$ |
|    Pressure variation | |
|    Oscillation of keyhole | |
| Melting front | $<10^{-3}$ |
|    Motion and heating of surrounding liquid | |
|    Thermal time constant | |
| Penetration depth | $<10^{-2}$ |
|    Variation in final penetration of weld | |

**TABLE 9.5. Time Scales for Signal Changes Related to Changes in Specific Weld Conditions**

| Condition | Time Scale |
|---|---|
| Gap, gradual opening | $w/\phi v$ |
| Gap, sudden | $w/v$ |
| Notch | $w/v;\ L > 2w$ |
| | $L/v;\ L < 2w$ |
| Drift off seam | $g/v_P$ |
| Defocus | $\sim \dfrac{0.025w}{v_z \tan \theta}$ |

w, beam radius; v, scan speed; L, notch width; $v_P$, perpendicular drift speed; $V_z$, drift speed along the normal to the weld plane; $\theta$, angle of the incident beam to the normal to the plane; g, gap width; $\phi$, angle of the gap in radians.

As an example, with w = 0.2 mm and v = 6 m/min, a sudden gap in the weld seam will cause a change in signal level over a timescale $\Delta t \sim 2 \times 10^{-5}$ sec. A defect with L = 0.1 and w = 0.02 mm would produce a transient with $\Delta t \sim 2 \times 10^{-6}$ sec. Both these time scales are comparable to those associated with plasma fluctuations (Table 9.4).

## 9.3  MONITORING CONFIGURATIONS

There are a variety of physical processes that occur during laser welding that are sensitive to the laser-weld interaction and therefore can be used in the establishment of a closed loop for process control. A useful summary of dependent process parameters is given in Figure 9.2. Measurement of some of these parameters is relatively straightforward (e.g., plasma intensity), but other quantities, such as the depth of the weld pool, are not directly measurable except through the effect they have on other parameters (Muller et al. 1997). Some quantities, such as plasma refractive index, are not suitable variables for process control because they integrate over pathlength and are susceptible to interference from a variety of ambient sources.

The light, heat, and acoustic radiation produced during laser welding are freely emitted into the ambient medium and therefore can be detected at various locations. In practice, mechanical and system motion constraints limit the placement of sensors, whereas the view that these sensors have of the laser-weld interaction can restrict the information received. For example, an optical sensor that looks along the laser beam axis will observe the effects of both keyhole and plasma, whereas a sensor looking transverse to the laser beam will detect only emission from the plasma.

Even acoustic sensors are sensitive to this effect, and they may record different frequency components when oriented to collect sound from different locations.

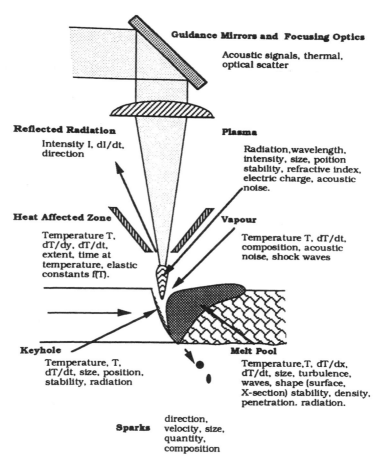

*Figure 9.2.* The main dependent parameters and the potential sources of process diagnostic signals. From Steen (1992).

Figure 9.3 shows fast Fourier transform (FFT) spectra of acoustic emission from lap welds at various processing speeds with the microphone located either near the laser head or under the sheets. It is apparent that the relative intensity of low-frequency components is enhanced in position 3. This has been shown to result in enhanced classification accuracy for welds under certain conditions (Gu and Duley 1996b).

The configuration shown in Figure 9.4 has been used by Miyamoto and Mori (1995) to monitor penetration depth during $CO_2$ laser welding of steel; it shows the time dependence of the output from optical detectors oriented 70° apart and reduced to show plasma ($P_p$) and keyhole ($P_k$) signals. Miyamoto and Mori (1995) found that $P_p$ and $P_k$ are in phase during penetration welding but are out of phase under partial penetration conditions. The sum of $P_p$ and $P_k$ is a convenient diagnostic tool for weld penetration.

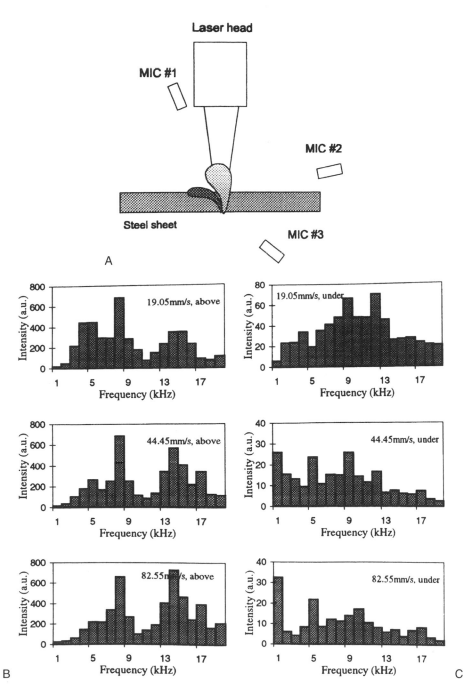

**Figure 9.3.** (a) Location of acoustic detectors; (b) acoustic spectra of position 1; (c) acoustic spectra at position 3. Lap weld, 1-mm steel sheets for 1.65-kW CO₂ laser. From Gu and Duley (1996b).

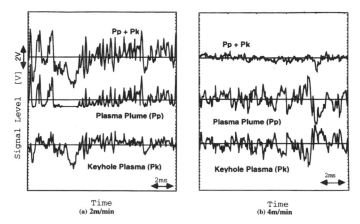

**Figure 9.4.** *Plasma and keyhole components of the signal in (a) full- and (b) partial-penetration welding. From Miyamoto and Mori (1995).*

Penetration depth also can be monitored by looking down the laser beam into the keyhole (Beyer et al. 1991, Maischner et al. 1991, Zimmerman 1993, Beersiek et al. 1997) using a scraper mirror to direct light from the keyhole back into a detector (Figure 9.5). A similar approach to imaging of plasma emission through the fiber optic has been shown to be successful in Nd:YAG laser welding (Haran et al. 1996, Morgan et al. 1996). Analysis of the frequency components associated with this light and the overall signal level has been shown to track weld penetration depth.

This system can provide a clear indication of joint gap and fit up problems (Beyer et al. 1992) through changes in signal level. An example of signal level changes encountered in the presence of a gap between the sheets in a butt weld is shown in Figure 9.5. This system is attractive because it is simple and can be configured as a simple set-point controller to detect alignment faults.

The same system has been demonstrated to be effective in the detection of holes and pits in the weld seam (Heyn, Decker, and Wohlfahrt 1994). In this case, both UV and near-infrared (NIR) sensors were used to detect keyhole conditions, with the NIR channel recording the presence of a melt while the UV photodiode detected plasma. Various correlations observed between these two signals were similar to those detected with sensors at angles off the optic axis (Chen et al. 1991, 1993). Specific weld defects can be recognized in the individual temporal signal components, but unambiguous assignment to the type of defect and whether it represents a fault condition (e.g., a through hole or a pit) usually require a comparison of the behavior of both UV and NIR signals. This behavior can be seen in the traces of VIS, UV, and NIR signals shown in Figure 9.6, which demonstrates that plasma collapse can occur without the disappearance of the melt. A transform of these temporal components to frequency space would provide additional diagnostic information.

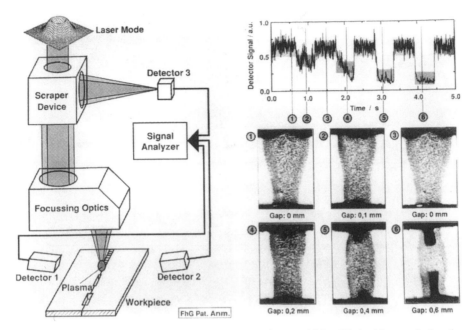

**Figure 9.5.** *Detection of the gap in butt welding of 3-mm-thick mild steel by monitoring the plasma light. From Beyer et al. (1992).*

A combination of two IR sensors, each looking at points on the weld pool, was used by Houlot and Nava-Rüdiger (1995) for process sensing and control. The sum and difference of the signals from these two detectors were found to be diagnostic of misalignment and fit up problems in butt welds (Figure 9.7). This is a good example of the way in which a simple level control can be implemented to obtain relative sophisticated information about the state of the welding process. Such a system also can be fast because data processing can be kept to a minimum.

Pyrometric temperature measurement with a PID controller was described by Deinzer et al. (1995) in an application involving folded seam welding of automobile components. The response time of this pyrometer was <2 msec, enabling real-time control over laser power to optimize well penetration and bead quality. A combination of IR and UV with reflected laser power has been shown to be effective in quality control of Nd:YAG spot welds (Griebsch et al. 1996).

Visual imaging of the laser weld pool and surface morphology is possible with a CCD video system (Li et al. 1993, Kinsman and Duley 1993, Beersiek et al. 1997). Filtering of the light emitted is essential to remove interference from the surface plasma. Images typically reveal the geometry of the melt pool and surface morphology, although the slow framing rate of conventional video cameras (30 Hz) prevents the detection of rapidly varying features. Video images may be enhanced with the use of image processing to delineate particular structures. Synchronization of the

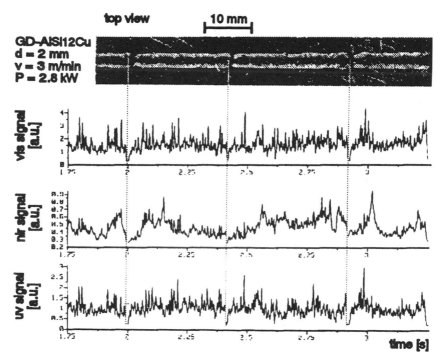

**Figure 9.6.** Welding of die-cast Al showing detection of welding defects by continuous measurement of the light emission of the process. From Heyn et al. (1994).

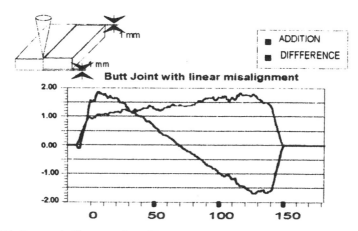

**Figure 9.7.** Sum and difference of two IR sensor outputs in laser welding of a butt seam with a surface mismatch. From Houlot and Nava-Rüdiger (1995).

framing rate with a secondary source of illumination can provide more detail on surface structure.

The large amount of data generated in imaging together with the need for signal processing to extract process parameters may slow the overall sampling rate to much less than 30 Hz. Restriction of analysis to only those parts of the image deemed to be significant, together with parallel image processing in multiple processors, are ways to speed up the overall rate of information updating. The application of these techniques in the system described by Li et al. (1993) yielded an overall data display rate of ~3 Hz. This is more suitable in a supervisory capacity than in adaptive control, although adaptive control of conduction welding has been shown to be possible in a simple system (Kinsman and Duley 1993).

A comparison of video images of the top and bottom of a butt weld during processing was shown to be a useful way of detecting weld conditions and gap separation (Bagger et al. 1992); an example for $CO_2$ laser welding of Zn-coated steel is shown in Figure 9.8. These data were extracted from the video images via digital image processing. Similar signals can be obtained using photodiodes (Miyamoto and Mori 1995), but CCD images can provide better discrimination

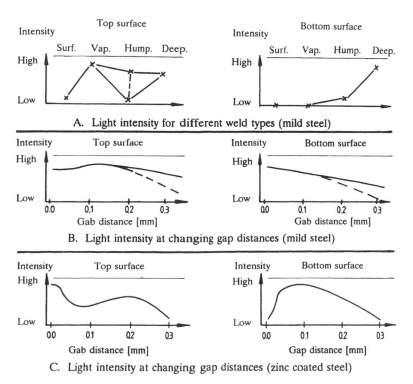

A. Light intensity for different weld types (mild steel)

B. Light intensity at changing gap distances (mild steel)

C. Light intensity at changing gap distances (zinc coated steel)

**Figure 9.8.** *General behavior of total light emission from laser welding process. Investigations were made from video images through digital image processing. From Bagger et al. (1992).*

against background noise because of the possibility of spatial resolution in the detected signal.

High-speed video imaging (Bagger et al. 1992, Griebsch et al. 1994a, Beersiek et al. 1997) provides the capability for detection of transient events and may be more suited to process control. With a 400-Hz framing rate, Bagger et al. (1992) were able to detect the rapid changes in weld pool width and area corresponding to a humping bead condition (Figure 9.9) during the welding of mild steel with 3-kW $CO_2$ laser radiation. Extraction of these data from the video image yielded an effective time resolution of <10 msec.

Video imaging at even higher framing rates (1000 Hz) was used by Griebsch et al. (1994a) to investigate weld pool dynamics during pulsed Nd:YAG and $CO_2$ seam welds. A frequency-doubled Nd:YAG laser was used as the illumination source. The spatial detail available with this technique is impressive and provides a high level of supervisory control over the laser welding condition.

A defect-sensing technique based on plasma charge detection was developed by Li et al. (1990), Shi et al. (1992), and Steen (1992). A variety of configurations are possible (see Figure 8.20), but all involve measurement of the potential of a point above the surface or the potential of the laser beam gas nozzle relative to the workpiece. In the latter case, the nozzle is insulated from the workpiece and allowed to float to a higher potential. The signal from the plasma sensor is related to charge generation in the plasma over the workpiece and thus tracks the presence of this plasma. As a result, deviations from a steady state solution are readily detected (Figure 9.10). Because the system measures charge directly, the response time can be fast. Li et al. (1990) found that the response is likely limited by the flight time of electrons or negative ions to 25–40 $\mu$sec. Under certain conditions, photoelectric effects may become important. The advantages of plasma probes for measurement

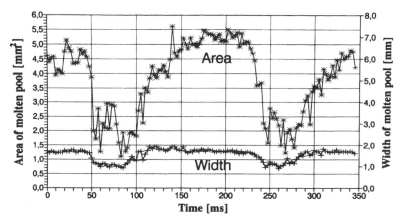

**Figure 9.9.** *Area and width of humping bead molten pool, observed with 400-Hz high-speed video camera. From Bagger et al. (1992).*

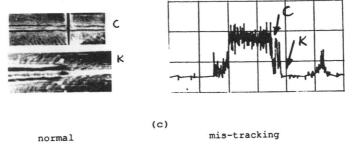

(c)

normal                    mis-tracking

**Figure 9.10.** *Sensor response to weld fault. From Li et al. (1990).*

of the presence of the welding plasma include a fast response time, simple robust construction, good signal-to-noise ratio, and low cost.

A comparison of different weld monitoring techniques by Watanabe et al. (1995) has shown that the combination of signals from different sources and types of detectors may be the most reliable way of detecting fault conditions. Signals obtained from acoustic, plasma spectral, and plasma potential detectors were correlated with CCD video images. Preliminary data suggested that common weld defects could be recognized via changes in signal level in each of these detection modes. An application in the diagnosis of faults in laser welding of cars was reported by Shi et al. (1992).

The monitoring of pulsed laser welding for quality control and possible closed-loop applications was described by Griebsch et al. (1995, 1996), Habenicht, Stark, and Deimann (1991), and Jurca et al. (1994). The results of these experiments show that fault conditions can be recognized for both spot and seam welding. The method described by Jurca et al. (1994) uses photodiode detectors to measure plasma (P), spatter (S), and temperature (T). The system is taught to relate P, S, and T signal characteristics to welding conditions and then can be set to an automatic mode that detects weld defects and changes in welding conditions. Failure processing is carried out using fuzzy logic routines.

The approach adopted by Habenicht, Stark, and Deimann (1991) to diagnose faults in spot welding uses FFT analysis to extract frequency components from acoustic emission. These are related to spiking frequencies in the laser pulse and to those in optical spatter and plasma emission. Acoustic emission during solidification was recorded and could be used to detect cold cracking. Unfortunately, no correlation was observed between acoustic emission and joint strength.

Griebsch et al. (1995) used the time-dependent reflectivity of the keyhole to follow the welding process. The time-dependent reflected power is found to be quite different for ''good'' and ''bad'' seam qualities (Figure 9.11). Workpiece heating, keyhole formation, and deep welding conditions were identifiable from characteristics of the reflected signal at the laser wavelength observed back along the axis of the laser beam.

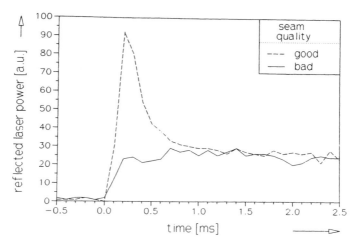

**Figure 9.11.** Signal spectrum of the reflected laser power and correlated seam qualities, $P_{peak}$ = 1500 W. From Griebsch et al. (1995).

## 9.4 CONTROL VARIABLES

Ultimately, control of laser welding involves the management of some aspect of the time history of heat input to the workpiece. This can be accomplished through the adjustment of incident laser intensity and its location on the workpiece. The heat input to the workpiece depends on the response of the workpiece to this irradiation and will be a function of focussing conditions, beam placement on the surface, and dwell time at the point of irradiation. There are, therefore, a limited range of control variables (Table 9.6).

With continuous wave lasers a common approach has been to modify the laser power in response to changes in welding condition so as to return the system to an optimized state. This method is the basis for the plasma shielding control (PSC) system developed by Seidel et al. (1993) and works well with radiofrequency-excited $CO_2$ lasers because of their high pulse repetition rate and short pulse duration. With this technique, the plasma over the workpiece is monitored, and when the signal exceeds a particular level, the laser operation is terminated for one or more pulses until the plasma signal dissipates. The ability to respond to changes in signal level over time scales of <1 msec makes this a true adaptive control system. For example, with a welding speed of 12 m/min, as encountered in tailor blanking operations, an interrupt time of 1 msec corresponds to a linear dimension of ~0.2 mm, or the width of the laser beam at the focus. This means that in principle, weld faults can be diagnosed and the system response can be modified to inhibit the formation of pits and holes for real-time quality control.

Stabilization of laser output also plays a role in the effective control of laser welding, although fluctuations are always present during laser welding due to the chaotic nature of the interaction between the laser beam and the workpiece (Otto,

Geisel, and Geiger 1996). Fluctuations in laser power are particularly important when welding reflective metals such as Al, for which feedback of laser radiation into the laser cavity can result in power instabilities. This problem was addressed by Deinzer et al. (1995). They describe a feedback system with a PI(D) controller that takes the output from a pyroelectric detector to correct the laser output for fluctuations and instabilities. The transient response of the system was greatly enhanced through this feedback loop. Improvements in $CO_2$ laser welding of Al alloy was demonstrated.

A more sophisticated control environment that combines both laser stabilization with an adaptive optical system and adjustment of workpiece motion was described by Otto, Deinzer, and Geiger (1994). A schematic of the elements of this system is given in Figure 9.12. This system was optimized to accommodate transients associated with the startup and termination of laser welding and to reduce problems that occur when the welding trajectory must be changed during welding of curves and contours in three-dimensional parts. The effect of optimization of motion control and laser intensity during the start of welding of a seam is shown in Figure 9.13; the objective was to find a set of operating conditions that enabled a smooth transition into a stable welding condition from a state of zero welding speed.

Stabilization of welding can also be accomplished through incorporation of a modulating term in the laser power Otto, Deinzer, and Geiger (1994) and Otto, Geisel, and Geiger (1996). This appears to work through resonant stabilization of the keyhole (Klein, Vicanek, and Simon 1996, Gu and Duley 1995). Otto, Geisel,

**TABLE 9.6. Primary Variables in the Control of Laser Welding**

CW laser
  Intensity, I(r,t)
    Absolute value
    Spatial distribution
      Focus
      Mode
      Position
    Temporal distribution
      Modulation
      Change in amplitude
Pulsed laser
  Intensity (as for CW)
  Pulse energy
  Pulse duration
  Repetition rate
  Pulse shape
Mechanical
  Scan speed
  Acceleration
Other
  Gas flow
  Gas composition

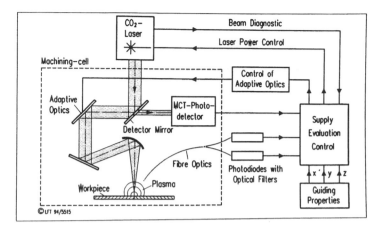

**Figure 9.12.** *Diagnostics and modular control system schematic. From Otto et al. (1994).*

and Geiger (1996) showed that a modulation term at 3.5 kHz greatly improves the uniformity of weld penetration depth. They suggest that the beneficial effect of modulating beam intensity occurs because it produces a constant absorbed power in the workpiece. However, the effect occurs only at certain modulation frequencies (~3.5 kHz when welding steel), which suggests a resonant interaction with a nonlinear dynamical system. Some theoretical aspects of this interaction are discussed by Otto, Geisel, and Geiger (1996) and Schulz et al. (1996).

Focal position also is an important parameter in any laser welding application, and the maintenance of the correct conditions through adaptive control is a key component in optimization of weld quality. This can be accomplished with an adaptive optic in the focussing system where the focal length is changed or by the motion

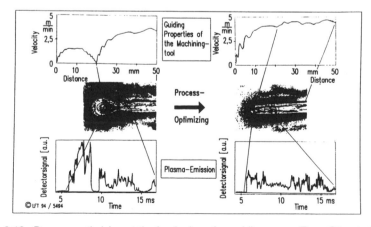

**Figure 9.13.** *Process optimizing at the beginning of a welding seam. From Otto et al. (1994).*

of the focussing element as a whole. Because the adaptive optic can be modulated over a wider band width, it is more suited to control under conditions in which rapid changes in focus must be accommodated. Motion of the optic as a whole is more difficult and results in slow response times, but it may be adequate in many practical applications, such as tailor blanking.

Optimization of the focal length in a practical laser welding system through closed-loop control with the use of a neural net approach was described by Bagger, Gong, and Olsen (1994). A schematic of this system is shown in Figure 9.14. The neural network was trained through a series of experiments in which the focal point was moved in a progressive way from a point above the workpiece to below it. Emission from the weld was recorded using photodiodes mounted above and below the workpiece during training, but only the upper diode was used during subsequent control (Figure 9.14). This system was able to identify and access the optimum focal point with an average error of 0.18 mm at a standard deviation of 0.36 mm.

Haran et al. (1996) developed a focus control system for Nd:YAG welding through analysis of the chromatic aberration of light from the plasma returned through the fiber-optic beam delivery system. Closed-loop control of focal position was demonstrated in welding with a 2-kW laser. The bandwidth of the mechanical system was 70 Hz, whereas that of the detection system was 100 Hz.

An adaptive control system based on the application of fuzzy logic to the analysis of aspects of a CCD image obtained during welding was reported by Kinsman and Duley (1993). A schematic showing the primary elements of this system is shown in Figure 9.15. The philosophy adopted in this system was to relate the number of bright pixels within certain regions of the weld image and to use this information to update the weld speed. Laser power and focal conditions both were kept constant in this case. Image data in $64 \times 64$–pixel areas, defined as number of bright pixels

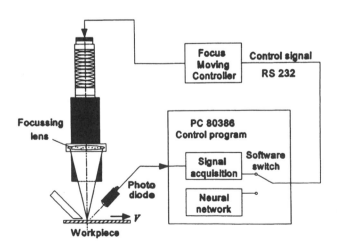

**Figure 9.14.** Principle of closed-loop control system based on photodiode monitoring and neural networks. From Bagger et al. (1994).

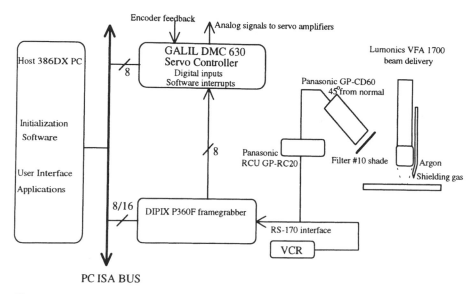

Encoder feedback

Analog signals to servo amplifiers

Host 386DX PC

GALIL DMC 630
Servo Controller
Digital inputs
Software interrupts

Lumonics VFA 1700
beam delivery

Panasonic GP-CD60
45°from normal

Initialization
Software

Panasonic
RCU GP-RC20

Filter #10 shade

Argon

Shielding gas

User Interface
Applications

DIPIX P360F framegrabber

RS-170 interface

VCR

PC ISA BUS

**Figure 9.15.** *Schematic showing the system components and data pathways for the prototype $CO_2$ expert welding system. From Kinsman and Duley (1993).*

(#BP) and change in the number of bright pixels (ΔBP), were obtained after preprocessing with an image processor. These were fuzzified and applied to a set of membership functions (Figure 9.16) in an inference engine to determine the degree to which they are "large" or "small." Values of #BP and ΔBP resulting from the fuzzification process correspond to antecedents in the rule base, which are then ANDED together using a minimum function to form the strength of the rule. This strength is applied to each of the rule outputs or consequents. If the output has already been assigned a rule strength during the current inference pass, a maximum function is used to determine which strength should apply. The defuzzification process completes the mapping from the input fuzzy sets to the output fuzzy set of weld speeds. The application of a center-of-gravity algorithm to obtain a weighted average of the triggered output membership functions avoids problems involving possible conflicting actions on the welding speed to be triggered. A 10-frame running average was used to reduce noise. Figure 9.17 shows implementation of this control system in the acquisition of a stable welding state in this prototype system (Kinsman and Duley 1993).

A study of real-time control using CCD imaging with other sensors was described by Derouet, Caillibottle, and Kechemair (1994). The general scheme adopted is shown in Figure 9.18. By using a filter in front of the CCD camera, IR wavelengths could be selected to permit detection of the location of the hottest point on the workpiece, the temperature at this location, and the length of the weld pool. Extraction of the melt pool depth was shown to be possible using an algorithm applied to analyze the digital video images. Although the control loop was not closed in these

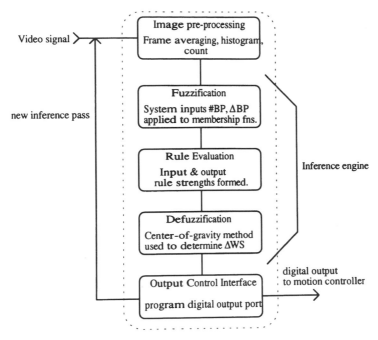

**Figure 9.16.** *Schematic of data flow in the fuzzy welding controller. From Kinsman and Duley (1993).*

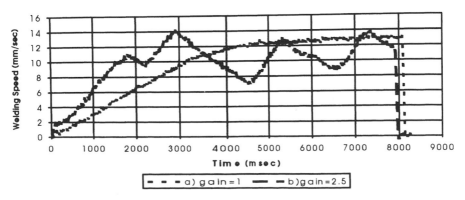

**Figure 9.17.** *Velocity profiles of bead on plate penetration welds under FL control. $CO_2$ laser power was 1500 W with transverse Ar shielding gas at 20 liter/min. Weld speed was given an initial value of ~1 mm/sec. a) Profile indicates a slow but stable increase in welding speed until good penetration is achieved at ~13 mm/sec. b) Relative gain of the system increased by 2.5×, resulting in a quicker but unstable increase in welding speed. The oscillations indicate the system is underdamped. From Kinsman and Duley (1993).*

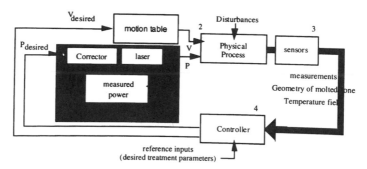

**Figure 9.18.** *Feedback control system schematic using thermal field. From Derouet et al. (1994).*

preliminary experiments, the results obtained suggest that regulation of the depth of the weld pool may be possible using data extracted from CCD images.

Another investigation of thermal imaging for control of laser welding was reported by Ocana et al. (1994). This approach was model based with the aim of developing an expert knowledge base. The inverse of the operating speed was adopted as the control variable because it was found to have a quasilinear relationship to the width of the welding seam. Simulations indicated that such a monitoring and control philosophy would lead to effective control.

A general analysis of the use of thermal imaging for control of laser welding was provided by Rajendran and Pate (1988). Further data on IR emission during welding are given by Semak, McCay, and McCay (1993) and Griebsch et al. (1996).

## 9.5   INTELLIGENT CONTROL

Monitoring of sensor outputs for changes in signal level or other signal characteristics is the primary activity of any control system. The outputs from these sensors after conditioning can be fed into one or more control loops that act to change process variables so as to maintain operation at optimized locations on the control surface. Operating points are located after a series of trial experiments in which the result of laser welding under the given condition is evaluated. Such closed-loop operating systems are capable of maintaining optimized welding but often are inflexible and do not always respond to changes in an intelligent manner.

A more flexible approach is to superimpose a diagnostic level that evaluates the sensor outputs in an intelligent way and provides supervisory control. This diagnostic level contains a separate feedback loop that compares observed inputs with a database and optimizes the output response. One logical framework for such an intelligent control system is shown schematically in Figure 9.19 (Steen 1992).

Decision making within the diagnostic level may involve setting up of a probability matrix that relates input variables in a causal manner to system parameters. A simple example of this approach for laser welding is given in Table 9.7. Protocols

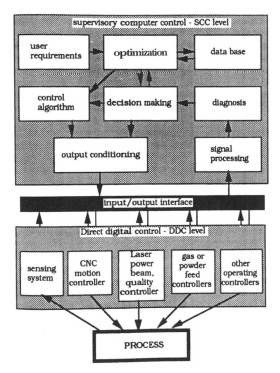

**Figure 9.19.** *Diagram of the logical framework for an "intelligent" control system. From Steen (1991).*

for assigning probabilities and for analysis of the sensor signals are derived from test experiments using the general philosophy described by Li et al. (1992).

In physical systems in which complex, ill-defined processes are encountered and the knowledge available often is imperfect or incomplete, information processing is best carried out using an artificial neural network to classify and evaluate the information available. Such fuzzy or imprecise information contains details about the behavior of the system at hand, but this information is not compatible with

**TABLE 9.7. Examples of Possible Signal Input Variables, Diagnostic Response, and Output Responses for Butt Welding**

| Signal (s) | Diagnosis (d) | Response (r) |
|---|---|---|
| Narrow weld bead | Speed too slow | Change laser power |
| Incomplete penetration | Speed too high | Change weld speed |
| Spatter | Laser power too low | Adjust flow |
| Overheated weld | Laser power too high | Gas flow |
| | Out of focus | |
| | Incorrect gas flow | |

specific mathematical algorithms relating input and output variables. This situation is encountered frequently during laser materials processing, particularly in laser welding.

Some advantages of neural networks were identified by Farson, Fang, and Kern (1991) and Shaw (1994), including:

1. ability of network to generalize principles from incomplete data
2. dealing with complex classification tasks
3. fast computation
4. relatively low cost
5. elimination of need for procedural programming or knowledge from experts or end users

Standard methodologies for designing fuzzy logic controllers, on the other hand, often are not suitable for laser welding, and successful implementation requires fine tuning and experimentation.

The neural network is based on an architecture that contains one or more hidden layers (Figure 9.20). Each input is connected to each ''neuron'' in the hidden layer, but these connections are weighted via a connection strength. The number of neurons

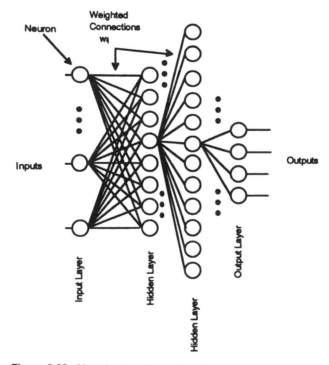

**Figure 9.20.** *Neural network structure. From Farson et al. (1991).*

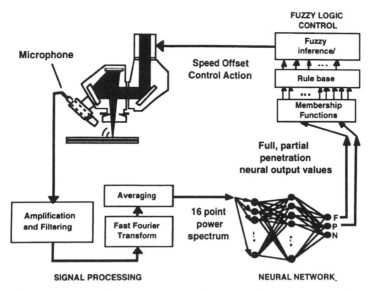

**Figure 9.21.** *Block diagram of an air-borne acoustic laser weld penetration control system. From Farson et al. (1994).*

in the input layer is determined by the number of input variables, whereas that in the output layer is equal to the number of output variables. As an example, the input variables could include the filtered output of an optical or acoustic signal measured from the workpiece after FFT processing, with individual spectral components assigned to specific input neurons. The output variable is typically weld penetration depth. This output then can be used to derive a control system based on fuzzy logic principles that outputs to the motion control system or to laser power. Evaluation of the output variable by the fuzzy logic controller results in the optimal change in weld speed or laser power to maintain correct welding conditions. A schematic of such a system architecture is shown in Figure 9.21. In this case, the input to this system is obtained from an acoustic monitor, but it also could involve plasma (Shi et al. 1992), optical (Beyer et al. 1994), or CCD signals (Kinsman and Duley 1993).

The connection strengths between neurons are obtained by a ''back-propagation'' technique, by which the input layer is given a specific set of inputs (i.e., a pattern) and yields an output that is compared with the known result.

The difference between the predicted output and the known result is used to assign an error. The connection strengths are adjusted to minimize this error. This procedure is repeated many times to optimize the values for the connection strengths and to minimize the error between true and predicted output variables. Iteration is continued until the error is within an acceptable range. A substantial training program may be necessary: for example, Beyer et al. (1994) found that classification of full penetration welds with up to 98.5% accuracy was possible only after training of the neural net with 60 different welds. This system also detected gaps of >0.2 mm with 99% probability in a sample of 300 welds.

## 9.6   SEAM TRACKING

The ability of a welding system to follow the seam between two components is crucial to successful welding operations. In principle, seam tracking is not required when the seam position in space is known, and this information can be programmed into a CNC system that moves the welding head along this trajectory. This generally is the case for simple linear welds in tailor blanks in which the position of the seam is carefully controlled and care is taken to maintain this geometry by clamping during welding. In many other practical applications of laser welding, however, the position of the seam either is not constant due to motion of the part during welding or cannot be predicted with sufficient accuracy. The extent of this problem is evident when one realizes that a drift of $\leq 0.2$ mm may be all that is required to go from satisfactory to unsatisfactory welding conditions.

Seam tracking can be carried out in two general ways. In the first method, some aspect of an acoustic, optical, or plasma signal that reflects seam geometry is monitored and used to control the position of the laser beam relative to the seam. The second technique uses a vision sensor to monitor the seam position either by projection of light from one or more auxiliary laser beams, or by CCD imaging of the workpiece together with data processing of the image to extract seam position.

With signal processing from an optical or acoustic sensor, the primary difficulty lies in assigning an observed change in signal to a specific directional change in the system (i.e., to distinguish "left" from "right"). Generally, this is not possible through monitoring of plasma or acoustic emission from a single detector even after the application of FFT pattern recognition algorithms. Exceptions to this rule may occur in the welding of dissimilar materials, for which spectral features attributable to specific elemental components may be distinguished. An example of this approach was discussed by Mueller et al. (1996) in a study of seam detection in butt welds between Cu and steel and mild/stainless steel.

Vision-based seam trackers were demonstrated in laser welding applications by Lucas and Smith (1988), Kuhnert (1993), Kreidler and Nitsch (1993), and Trunzer, Lindl, and Schwarz (1993). These devices yield seam accuracies as good as $\pm 0.1$ mm and have been adapted to robotic laser welding. Operation is at scanning rates of up to 50 Hz, permitting integration into laser welding systems at speeds in excess of 10 m/min. A simplified representation of such a system is shown schematically in Figure 9.22 (Trunzer, Lindl, and Schwarz 1993). Key variables include the lead distance (i.e., the distance between the laser focus and the point at which the seam is sampled). This determines the welding rate that can be accommodated and the contour critical angle. The contour critical angle is a measure of the smallest change in curvature of the seam that can be detected. This can be minimized through mounting of the sensor head onto a rotational mount, which allows it to track independently of the position of the laser focus. This capability is most useful in three-dimensional welding geometries (Figure 9.23).

Closed-loop control in laser welding of automotive components was reported by Tönshoff, Overmeyer, and Schumacker (1996). The general aspects of control structures for laser welding, including process planning, path planning, process simulation, and workpiece processing, are discussed by Kaierle et al. (1996).

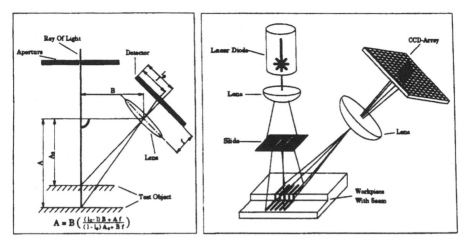

**Figure 9.22.** *Principle of triangulation and a schematic view of a multiple light section sensor. From Trunzer et al. (1993).*

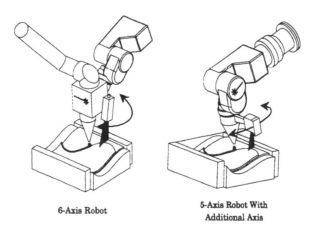

6-Axis Robot                5-Axis Robot With
                            Additional Axis

**Figure 9.23.** *Robots with external (left) and internal beam guidance system. From Trunzer et al. (1993).*

# Appendix

# Industrial Applications of Laser Welding: A Short Bibliography

Practical applications of laser welding are found in many industries. The following is an abbreviated list of journal articles that describe specific laser welding applications. These articles contain useful information on the implementation of laser welding technology in the manufacture of a variety of components and systems.

## Automotive—General

Plasma-augmented lasers shine ahead in autobody application.
Biffin, J. 1997. *Weld. Metal Fabrication* April: 18.
Lasers weld with high power.
Murray, M. 1995. *Am. Machinist* July: 45.
Laser welding of automotive and aero components.
Kunzig, L. 1994. *Weld. Metal Fabrication* Jan.: 14.
Lasers zapping spot welders off the auto line.
Vaccari, J. A. 1994. *Am. Machinist* March: 31.
YAG lasers strike with more power.
Vaccari, J. A. 1994. *Am. Machinist* May: 55.
Laser beam welding and GMA welding go on stream at Arvin Industries.
Irving, B. 1993. *Welding J.* 72:47.
Laser beam welding of high hardness steels: Applications to armoured vehicles.
Bourges, P., Berniolles, J. 1992. Proceedings ICALEO '92, 574.
Applications of laser materials processing in Toyota Motor Corporation.
Mikame, K. 1992. Proceedings LAMP '92, 947.
Laser welding of components for the automotive industry.
Klein, R., Fischer, R., Polzin, R., et al. 1992. Proceedings LAMP '92, 975.
Laser technology within the Volvo Car Corporation.

Hannicke, L. 1988. Proceedings 5th Lasers in Manufacturing, 97.
Task-oriented laser processing cells consisting of an industrial robot and a laser beam guiding system.
    Schraft, R. D., Hardock, G., Konig, M. 1988. Proceedings 5th Lasers in Manufacturing, 313.
Metalworking lasers in automobile fabrication—A view from the U.K.
    Dawes, C. J. 1986. Proceedings SPIE 668, 186.
Use of lasers in seam welding of engine parts for cars.
    Luttke. 1986. Proceedings SPIE 668, 193.
Applications of laser processing for automotive manufacturing in Japan.
    Ito, N., Ueda, K., Takagi, S. 1986. Proceedings SPIE 668, 201.
A five-axis robotic laser and vision integrated ''on-line'' welding system.
    Uddin, N., Berardi, E., DuCharme, R. C. et al. 1986. Proceedings SPIE 668, 260.

## Automotive—Gears

Production laser welding of gears.
    Guastaferri, D. 1986. Proceedings SPIE 621:40.
Production laser welding of gears.
    Guastaferri, D. 1986. Proceedings SPIE 668:223.
Industrial applications of high power $CO_2$ lasers—System description.
    Gukelberger, A. 1986. Proceedings SPIE 650:254.

## Automotive—Laser-Welded Blanks

A cost comparison of weld technologies for tailor welded blanks.
    Baron, J. S. 1997. *Welding J.* 76:39.
Laser beam welding shifts into high gear.
    Irving, B. 1997. *Welding J.* 76:35.
Laser welded tailored blanks in the automotive industry.
    Waddell, W., Davies, G. M. 1995. *Weld. Metal Fabrication* March: 104.
Welded blanks tailor profits.
    Chalmers, R. E. 1994. Forming and Fabricating, Mar., p. 13.
Automated laser systems for high volume production sheet metal application.
    Betz, U., Retzbach, M., Alber, G., et al. 1992. Proceedings ICALEO '92, 627.
Blank welding forces auto makers to sit up and take notice.
    Irving, B. 1991. *Welding J.* 70:39.

## Cans

Development and implementation of high-speed laser welding in the can-making industry.
    Sharp, C. M. 1987. Proceedings LAMP '87, 541.

## Consumer Goods

Laser beam welding goes into high-speed production of home hot water tanks.
 Garrison, G. 1993. *Welding J.* 72:53.
Lasers deliver precise welds on new razor.
 1990. *Welding J.* 69:47.
Laser welding of TV picture tubes.
 Klinkenberg, E. G. H. 1989. Proceedings 6th Lasers in Manufacturing, 155.
High-power $CO_2$ laser welding of food mixer parts: A case history.
 Notenboom, G. J. A. M., Jelmorini, G. 1986. Proceedings SPIE 650, 295.

## Cylinders, Tubes, Pipes

Development of high-power laser welding process for pipe.
 Hayashi, T., Inaba, Y., Matuhiro, Y., et al. 1996. Proceedings ICALEO '96, D132.
Feasibility evaluations for the integration of laser butt welding of tubes in industrial pipe coil production lines.
 Penasa, M., Columbo, E., Giolfo, M. 1994. Proceedings SPIE 2207, 200.
Laser beam welder for thinner pipe.
 Tang, Z., Dai, B., Feng, C., et al. 1992. Proceedings LAMP '92, 559.
Flexible manufacturing cell for $CO_2$ laser welding of tubular parts.
 Engstrom, H., Ilar, T., Nillson, K. 1987. Proceedings 4th Lasers in Manufacturing, 107.
Girth welding of X-60 pipeline steel with a 10kw laser.
 Megaw, J. H. P. C., Hill, M., Osbourn, S. J. 1986. Proceedings SPIE 680, 311.
Laser welding of cylinders.
 Kraencke, R., Gregson, V. 1986. Proceedings SPIE 621, 2.

## Lamps

Implementation of laser welding for lamp leads.
 Doubrava, J. H., Ticknor, G. W., Jones, M. G. 1990. Proceedings ICALEO '90, 400.
Lasers in lamp making: Ten years experience in Thorn Lighting Ltd., U.K.
 Howe, S., Morris, R. 1987. Proceedings 4th Lasers in Manufacturing, 9.

## Nuclear

Laser beam cutting and welding creates precision, lightweight structure.
 Lund, O., Bartoszek, L. 1993. *Welding J.* 72:71.
High-power YAG laser welded sleeving technology for steam generator tubes in nuclear power plants.
 Ishide, T., Nagura, Y., Matsumoto, O., et al. 1992. Proceedings LAMP '92, 957.
Laser welding of sleeves for nuclear reactor plant heat exchanger tubes.
 Miller, R. A., Bruck, G. J., Morgen, E. P., et al. 1988. Proceedings ICALEO '88, 339.

## Packaging

Selective sealing and bonding of silicon-glass compounds and LIGA components by laser beam welding.

Gillner, A., Legewie, F., Poprawe, R. 1996. Proceedings ICALEO '96, E148.

A guide to welding with low-power YAG lasers.

Marley, C. 1996. *Welding J.* 75:47.

Laser soldering of Sn-plated brass integrator assembly housings.

Keicher, D. M., Poulter, G. A., Sorensen, N. R. 1993. Proceedings ICALEO '93, 721.

Laser beam welding of stainless steel tubes used for pure gas supplying in microchip, manufacturing.

Han, V. H., Roth, G. Ezel, S. 1988. Proceedings Lasers in Manufacturing, 215.

Spot and seam welding by solid state lasers used in conjunction with beam guidance systems including light cables.

Seiler, P. 1988. Proceedings 5th Lasers in Manufacturing, 157.

Laser welding in the assembly of a high reliability aluminum alloy amplifier housing.

Coyle, R. J., Solan, P. P., Martin, J. C. 1987. Proceedings ICALEO '87, 75.

Development of an automatic laser welding machine to produce gas-filled capsules.

Stevenson, P. 1986. Proceedings 3rd Lasers in Manufacturing, 157.

Laser welding systems for hermetic sealing.

Bosnos, C. M. 1986. Proceedings SPIE 668:310.

## Structures

Laser welder for the continuous cold rolling mill: Application to high carbon steels.

Fukuhara, A., Koyama, T., Komatsu, T., et al. 1992. Proceedings ICALEO '92, 584.

Laser welding of metal honeycomb panel with multiple reflecting effects of high-power laser beams.

Minamida, K., Oikawa, M., Sugihashi, A., et al. 1992. Proceedings ICALEO '92, 557.

Laser skid welding of T-joints for ship fabrication.

Brookes, S. J. 1988. Proceedings Lasers in Manufacturing, 165.

## Saw Blades

Laser welding of diamond saw blades.

Li, L., Jin, X., Shi, X. 1996. Proceedings SPIE 2888, 178.

The basics of laser welding.

Darchuk, J. M., Migliore, L. R. 1985. Lasers and Applications, Mar. 59.

# References

Abdullah, H.A., Ng, E.S., Chorwin, C.R., et al. 1995. Proceedings ICALEO '95, 923.

Abe, N., Agano, Y., Tsukamoto, M., et al. 1996. Proceedings ICALEO '96, D64.

Adair, R. 1994. *The Fabricator* 24:96.

Akhter, R., Davis, M., Dowden, J., et al. 1989a. *J. Phys. D. Appl. Phys.* 22:23.

Akhter, R., Steen, W.M. 1990. Proceedings ISATA '90, 219.

Akhter, R., Steen, W.M., Cruciani, D. 1988. Proceedings 5th Conference Lasers in Manufacturing, 195.

Akhter, R., Watkins, K.G., Steen, W.M. 1989b. Proceedings 6th International Conference Lasers in Manufacturing, 121.

Albright, C.E., Chiang, S. 1988a. Proceedings ICALEO '88, 207.

Albright, C.E., Chiang, S. 1988b. *J. Laser Appl.* Fall:18.

Albright, C.E., Hsu, C., Lund, R.O. 1990. Proceedings ICALEO '90, 357.

Anderson, J.E., Jackson, J.E. 1965. *Weld. J.* 44:1018.

Andrews, J.G., Atthey, D.R. 1976. *J. Phys. D.: Appl. Phys.* 9:2181.

Anisimov, S.E., Bonch-Bruerich, A.M., Elyasherich, A.M. 1967. *Sov. Phys. - Tech. Phys.* 11:945.

Aoh, J.N., Kuo, F.H., Li, Y.M., et al. 1992. International Trends in Welding Science & Technology '92, 649.

Arai, N., Okita, K., Aritoshi, M., et al. 1987. Proceedings LAMP '87, 221.

Arata, Y. 1987. Proceedings LAMP, 3.

Arata, Y., Maruo, H., Miyamoto, I., et al. 1986. Proceedings Conference Electron Laser Beam Welding, Inst. Weld., 159.

Aruga, S., Matsui, E., Okino, K., et al. 1992. Proceedings LAMP '92, 517.

Arzuov, M.I., Barchukov, A.I., Bunkin, F.N., et al. 1979. *Sov. J. Quant. Electron* 9:1308.

Ashby, M.F., Easterling, K.E. 1984. *Acta. Met.* 32:1935.

Aubert, P., Bouilly, P., Garcoin, J.Y., et al. 1987. Proceedings LAMP '87, 81.

Austin, P.D. 1986. Proceedings SPIE 668, 232.

Avilov, V.V., Vicanek, M., Simon, G. 1996. *J. Phys. D. Appl. Phys.* 29:1146.

Avramchenko, P.F., Molchan, I.V. 1983. *Auto. Weld.* May:68.

Baardsen, E.L., Schmatz, D.J., Bisaro, R.E. 1973. *Weld. J.* 52:227.

Baeslack, W.A., Cieslak, M.J., Headley, T.J. 1989. Recent Trends in Welding Science and Technology TWR '89, 211.

Bagger, C., Gong, H., Olsen, F. 1994. Proceedings SPIE 2207, 369.

Bagger, C., Miyamoto, I., Olsen, F., et al. 1992. Proceedings LAMP '92, 553.

Bagger, C., Nielsen, K.A., Olsen, F. 1993. Proceedings ICALEO '93, 693.

Bahun, C.J., Enquist, R.D. 1963. Proceedings Natl. Electron Conf. 18:607.

Ball, W.C., Banas, C.M. 1974. Welding with a High Power $CO_2$ Laser, Nat. Aerosp. Eng. and Mfr. Meeting, San Diego, California.

Banas, C.M. 1971. UARL Report 125 (June 1971), IEEE Symp. Electron Ion, Laser Beam Technology 11th, San Francisco Press, San Francisco.

Banishev, A.F., Golubev, V.S., Novikov, M.M., et al. 1994. Proceedings SPIE 2257:14.

Baranov, M.S., Metarhop, L.A., Geinrikhs, I.N. 1968. Weld. Prod. GB, 15, 23.

Barchukov, A.I., Bunkin, F.V., Konov, V.I., et al. 1975. *Sov. Phys. JETP* 39:469.

Basu, B., Date, A.W. 1990a. *Int. J. Heat Mass Transfer* 33:1149.

Basu, B., Date, A.W. 1990b. *Int. J. Heat Mass Transfer* 33:1165.

Baysore, J.K., Williamson, M.S., Adonyi, Y., et al. 1995. *Weld. J.* 74:345s.

Bea, M., Giesen, A., Hügel, H. 1993. Proceedings ISATA '93, 223.

Bea, M., Glumann, C., Grunewald, B., et al. 1994. Proceedings SPIE 2207, 111.

Beck, M., Berger, P., Hugel, H. 1992. Proceedings ECLAT, 693.

Behler, K., Beyer, E., Herziger, G., et al. 1988a. Proceedings ICALEO '88, 98.

Behler, K., Beyer, E., Schafer, R. 1988b. Proceedings ICALEO '88, 249.

Behler, K., Beyer, E., Schulz, W., et al. 1988c. Proceedings 5th Int. Conf. Lasers in Manufacturing, 187.

Bell, D.E., Petrolonis, K., Howell, P.R. 1992. Proceedings International Trends in Welding Science and Technology TWR '92, 301.

Beersiek, J., Poprawe, R., Schulz, W., et al. 1997. Proceedings ICALEO '97, in press.

Berger, M. 1986. Proceedings SPIE 650, 141.

Berger, R.M., Chmelir, M., Reedy, H.J., et al. 1991. Proceedings SPIE, 1397, 611.

Bermejo, D., Fabbro, R., Sabatier, L., et al. 1990. Proceedings SPIE 1279, 118.

Betz, U., Retzbach, M., Alber, G., et al. 1992. Proceedings ICALEO '92, 627.

Beyer, E., Abels, P., Drenker, A., et al. 1991. Proceedings ICALEO '91, SPIE vol. 1722, 133.

Beyer, E., Abels, P. 1992. LAMP, 433.

Beyer, E., Behler, K., Herziger, G. 1988. Proceedings LIM, 233.

Beyer, E., Brenner, B., Poprawe, R. 1996. Proceedings ICALEO '96, D157.

Beyer, E., Dilthey, U., Imhoff, R., et al. 1994. Proceedings ICALEO '94, 183.

Beyer, E., Gasser, A., Gatzweiler, W. 1987. Proceedings ICALEO '87, 17.

Beyer, E., Herziger, G., Kramer, R., et al. 1986. Proceedings ICALEO '86, 37.

Beyer, E., Imhoff, R., Neuenhahn, J. 1994. Proceedings IBEC '94, Advanced Technologies and Processes, 101.

Beyer, E., Maischner, D., Kratzisch, C. 1994b. Proceedings ICALEO '94, 51.

Bloehs, W., Dausinger, F. 1996. ICALEO '95, 1068.

Bobadilla, M., Lacaze, J., Lesoult, G. 1988. *J. Cryst. Growth* 89:531.

Bonello, L., Bailo, A. 1993. Proceedings ISATA '93, 119.

Borik, S., Giesen, A. 1988. *Proceedings Lasers in Manufacturing* 259.

Born, M., Wolf, E. 1975. Principles of Optics, Pergamon, New York.

Brandon, E.D. 1992. *Weld. J.* June: 55.

Bransch, H.N., Wang, Z.Y., Liu, J.T., et al. 1991. *J. Laser Appl.* 3:25.

Bransch, H.N., Weckman, D.C., Kerr, H.W. 1992. *Weld. J.* 71:11s.

Bruckner, M., Schafer, J.H., Uhlenbusch, J. 1989. *J. Appl. Phys.* 66:1326.

Carslaw, H.S., Jaeger, J.C. 1959. Conduction of Heat in Solids. 2nd ed., Oxford University Press, London.

Carlson, R.W., Gregson, V.G. 1986. Proceedings ICALEO '86, 211.

Carlson, K.W., Gregson, V.G. 1988. Proceedings ICALEO '88, 223.

Carter, S., Guastaferri, D. 1996. Proceedings ICALEO '96, Auto, 21.

Chalmers, R.E. 1994. *Forming & Fabricating* Mar: 13.

Chan, C., Mazumder, J., Chen, M.M. 1984. *Met. Trans.* 15A:2175.

Chande, T., Mazumder, J. 1984. *J. Appl. Phys.* 56:1981.

Charles, P., Engel, T., Fontaine, J., et al. 1988. Proceedings 5th LIM, 223.

Charschan, S.S. 1972. Lasers in Industry, van Nostrand-Reinhold, Princeton, N.J.

Chen, H.B., Li, L., Brookfield, D.J., et al. 1993. *NDT & E International* 26:67.

Chen, H.B., Li, L., Brookfield, D.J., et al. 1991. Proceedings ICALEO '91, 113.

Chen, T.C., Kannatey-Asibu, E. 1994. Proceedings ICALEO '94, 668.

Chen, G., Roth, G., Maisenhalder, F. 1993. *Laser Optoelek.* 25:43.

Chennat, J.C., Albright, C.E. 1984. Proceedings ICALEO '84, 76.

Chiang, S., Albright, C.E. 1992. Proceedings ICALEO '92, 491.

Chryssolouris, G. 1991. Laser Machining Theory and Practice, Springer-Verlag, New York.

Cieslak, M.J., Fuerschbach, P.W. 1988. *Met. Trans.* 19B:319.

Cline, H.E., Anthony, T.R. 1977. *J. Appl. Phys.* 48:3895.

Cohen, M.I. 1967. *Bell Lab. Rec.* 45:247.

Cohen, M.I., Epperson, J.P. 1968. *Advan. Electron. Electro. Phys. Suppl.* 4:139.

Cohen, M.I., Mainwaring, F.J., Melone, T.G. 1969. *Weld. J.* 48:191.

Colla, T.J., Vicanek, M., Simon, G. 1994. *J. Phys. D.: Appl. Phys.* 27:2035.

Collur, M.M., DebRoy, T. 1989. *Met. Trans.* 20B:277.

Collur, M.M., Paul, A., DebRoy, T. 1987. *Met. Trans.* 18B:733.

Copley, S.M., Beck, D.G., Esquivel, O., et al. 1981. Proceedings Lasers in Metallurgy, AIME, Mukherjee, K., Mazumder, J., eds. 11.

Corrodi, R. 1996. IBEC '96, Materials and Body Testing, 110.

Coyle, R.J. 1983. Proceedings LIMP, 185.

Dahotre, N.B., McCay, T.D., McCay, M.H., 1989. *J. Appl. Phys.* 65:5072.

Dahotre, N.B., McCay, T.D., McCay, M.H., 1993. Proceedings ICALEO '93, 498.

Dahotre, N.B., McCay, M.H., McCay, T.D., et al. 1990. Proceedings ICALEO '90, SPIE 1601:343.

Dahotre, N.B., McCay, M.H., McCay, T.D., et al. 1991. *J. Mater. Res.* 6:514.

Dausinger, F., Faisst, F., Glumann, C., et al. 1995. *Laser Optoelk.* 27:45.

Dausinger, F., Rapp, J., Beck, M., et al. 1996. *J. Laser Appl.* 8:285.

Dausinger, F., Rapp, J., Hohenberger, B., et al. 1997. Proceedings IBEC '97, Advanced Technologies and Processes, 38.

David, S.A., Vitek, J.M. 1981. Proceedings LIM, 247.

Dawes, C. 1992. Laser Welding, McGraw-Hill, New York (originally published by Abington Publishing, Cambridge, England).

DebRoy, T., Basu, S., Mundra, K. 1991. *J. Appl. Phys.* 70:1313.

Deinzer, G., Otto, A., Hoffmann, P., et al. 1995. *Manuf. Syst.* 24:111.

Dell'Erba, M., Sforza, P., Chita, G., et al. 1986. Proceedings ICALEO '86, 57.

Deminet, C. 1987. Proceedings ICALEO '87, 241.

Denney, P.E., Metzbower, E.A. 1989. Laser Materials Proceedings III, The Minerals, Metals and Materials Soc., Warrendale, Pennsylvania 83.

Denney, P.E., Metzbower, E.A. 1991. Proceedings ICALEO '91, SPIE 1722:84.

Derouet, H., Caillibottle, G., Kechemair, D. 1994. Proceedings SPIE, 2207:321.

Dilthey, U., Shu, X. 1993. *Weld. World* 31:36.

Di Pietro, F.A. 1992. Proceedings ICALEO '92, 621.

Douay, D., Daniere, F., Fabbro, R., et al. 1996. Proceedings ICALEO '96, D54.

Dowden, J. Davis, M., Kapadia, P. 1983. *J. Fluid Mech.* 126:123.

Dowden, J., Davis, M., Kapadia, P. 1985a. *J. Appl. Phys.* 57:4474.

Dowden, J., Davis, M., Kapadia, P. 1985b. *J. Phys. D.: Appl. Phys.* 18:1987.

Dowden, J., Kapadia, P., Postacioglu, N. 1989. *J. Phys. D.: Appl. Phys.* 22:741.

Dowden, J., Postacioglu, N., Davis, M., et al. 1987. *J. Phys. D.: Appl. Phys.* 20:36.

Ducharme, R., Kapadia, P., Dowden, J. 1992. Proceedings ICALEO '92, 188.

Ducharme, R., Kapadia, P., Dowden, J. M. 1993. Proceedings ICALEO '93, 177.

Ducharme, R., Kapadia, P., Dowden, J., et al. 1992. Proceedings ICALEO '92, 177.

Ducharme, R., Williams, K., Kapadia, P., et al. 1994. *J. Phys. D.: Appl. Phys.* 27:1619.

Duhamel, R.F. 1988. Proceedings ICALEO '88, 76.

Duhamel, R.F. 1996. Proceedings ICALEO '96, auto, 136.

Duhamel, R.F., Banas, C.M. 1983. Proceedings Lasers in Mat. 209.

Duley, W.W. 1976. $CO_2$ Lasers: Effects and Applications, Academic Press, New York.

Duley, W.W. 1983. Laser Processing and Analysis of Materials, Plenum Press, New York.

Duley, W.W. 1987. Proceedings LAMP '87, 13.

Duley, W.W. 1994. Proceedings IBEC '94, Automotive Body Materials, 50.

Duley, W.W. 1996. UV Lasers: Effects and Applications in Materials Science, Cambridge University Press, Cambridge.

Duley, W.W. 1986. Proceedings NATO ASI Laser Surface Treatment of Metals, Draper, C.W. Mazzoldi, P., eds. 3.

Duley, W.W., Gonsalves, J.N. 1972. *Can. Research and Development* Jan./Feb.: 25.

Duley, W.W., Mao, Y.L., Kinsman, G. 1991. Proceedings Conf. Laser and Electron Beam, 206.

Duley, W.W., Mao, Y.L. 1994. *J. Phys. D.: Appl. Phys.* 27:1379.

Duley, W.W., Mueller, R.E. 1992. *Polym. Sci. Eng.* 32:582.

Duley, W.W., Olfert, M., Bridger, P., et al. 1992. Proceedings LAMP '92, 261.

Dunlap, G.W., Williams, D.I. 1962. Proceedings Natl. Electron Conf. 18:601.

Dunn, I., Bridger, P.M., Duley, W.W. 1993. *J. Phys. D.* 26:1138.

Easterling, K. 1983. Introduction to the Physical Metallurgy of Welding, Butterworths, London.

Eberle, H.G., Richter, K., Schobbert, H. 1994. Proceedings SPIE 2207:185.

Engel, T., Fontaine, J. 1989. Proceedings SPIE 1132:282.

Enquist, R.D. 1962. *Met. Progr.* 82:67.

Essien, M., Fuerschbach, P.W. 1996. *Weld. J.* 75:47s.

Essien, M., Keicher, D.M., Jellison, J.L. 1995. Proceedings ICALEO '95, 583.

Exner, H., Gerber, B., Kimme, T., et al. 1993. Proceedings ICALEO '93, 611.

Faerber, M. 1997. Proceedings IBEC '97, Body Assembly and Manufacturing, 14.

Fairbanks, R.H., Adams, C.M. 1964. *Weld. J.* 43:97s.

Farson, D., Ali, A., Sang, Y. 1996. Proceedings ICALEO '96, auto, 72.

Farson, D.F., Fang, K.S., Kern, J. 1991. Proceedings ICALEO '91, SPIE 1722:104.

Farson, D., Hillsley, K., Sames, J., et al. 1994. Proceedings ICALEO '94, 86.

Farson, D., Sang, Y., Ali, A. 1997. *J. Laser Appl.* 9:87.

Fieret, J., Terry, M.J., Ward, B.A. 1986. Proceedings SPIE 668:53.

Finke, B.R., Kapadia, P.D., Dowden, J.M. 1990. *J. Phys. D.: Appl. Phys.* 643.

Finke, B.R., Simon, G. 1990. *J. Phys. D.: Appl. Phys.* 23:67.

Flavenot, J.F., Deville, J.P., Diboine, A., et al. 1993. *Weld. World* 31:12.

Forrest, M., Marttila, W.A., Tomakich, T.G., et al. 1997. Proceedings IBEC '97, Advanced Technologies and Processes, 69.

Fraser, F.W., Metzbower, E.A. 1983. Proceedings LIM '83, 196.

Fuerschbach, P.W. 1994. Proceedings ICALEO '94, 651.

Fuerschbach, P.W. 1996. *Weld. J.* 75:24s.

Fuerschbach, P.W., MacCallum, D.O. 1995. Proceedings ICALEO '95, 497.

Fukui, K., Uchihara, M., Takahashi, M., et al. 1996. Proceedings IBEC '96, Materials and Body Testing, 100.

Funk, M., Kohler, U., Behler, K., et al. 1989. Proceedings SPIE, 1132, 174.

Gagliano, F.P., Lumley, R.M., Watkins, L.S. 1969. Proceedings IEEE 57, 114.

Gagliano, F.P., Zaleckas, V.J. 1972. Lasers in Industry, Charschan, S.S., ed., Van Nostrand-Reinhold, Princeton, New Jersey, 139.

Garashchuk, V.P. et al. 1969. *Auto. Weld.* 2:56.

Garashchuk, V.P., Kirsei, V.I., Shinkarev, V.A. 1986. *Sov. J. Quant. Electron* 16:1660.

Garashchuk, V.P., Molchan, I.V. 1969. *Auto. Weld.* 9:12.

Garmine, E., McMahon, T., Bass, M. 1976. *Appl. Opt.* 15:145.

Gascoin, J.Y., Juguet, Y., Aubert, P., et al. 1987. Proceedings LAMP '87, 113.

Gatzweiler, W., Maischner, D., Faber, F.J., et al. 1989. Proceedings SPIE, 1132:157.

Geusic, J.E., Marcos, H.M., Vankitert, L.G. 1964. *Appl. Phys. Lett.* 4:182.

Giesen, A. 1992. Proceedings LAMP, Osaka, 213.

Gilath, I., Signamarcheix, J.M., Bensussan, P. 1994. *J. Matl. Sci.* 29:3358.

Gilgen, P., Kurz, W. 1996. Proceedings NATO ASI E307, 77.

Glumann, C., Rapp, J., Bea, M. et al. 1993a. Proceedings ISATA '93, 239.

Glumann, C., Rapp, J., Dansinger, F., et al. 1993b. Proceedings ICALEO '93, 672.

Gnanamuthu, D.S., Moores, R.J. 1987. Proceedings Conf. Laser Electron Beam, 295.

Gopinathan, S., Murthy, J., McCay, T.D., et al. 1993a. Proceedings ICALEO '93, 794.

Gopinathan, S., Murthy, J., McCay, T.D., et al. 1993b. Proceedings ICALEO '93, 661.

Gratzke, U., Kapadia, P.D., Dowden, J. 1991. *J. Phys. D.: Appl. Phys.* 24:2125.

Gregersen, O., Olsen, F.O. 1990. Proceedings ICALEO '90, SPIE 1601:28.

Grezev, A.N., Grigoryants, A.G., Fedorov, V.G., et al. 1984. *Auto. Weld.* Sept.: 29.

Griebsch, J., Berger, P., Dausinger, F., et al. 1994a. Proceedings SPIE 2246:136.

Griebsch, J., Berger, P., Dausinger, F., et al. 1994b. Proceedings ICALEO '94, 173.

Griebsch, J., Hugel, H., Dausinger, F., et al. 1995. Proceedings ICALEO '95, 603.

Griebsch, J., Schlichtermann, L., Jurca, M., et al. 1996. Proceedings ICALEO '96, B164.

Griem, H. 1974. Spectral Line Broadening by Plasmas, Academic Press, New York.

Grigoryants, G. 1994. Basics of Laser Material Processing, CRC Press, Boca Raton, Florida.

Grigoryants, A.G., Shiganov, I.N., Ivanov, A.M. et al. 1983. *Auto. Weld.* Sept: 24.

Grong, O. 1994. Metallurgical Modelling of Welding, Institute of Metals, London.

Gu, G., Morrow, C. 1994. Proceedings ICALEO '94, 803.

Gu, H. 1995. Masters Thesis, University of Waterloo.

Gu, H., Duley, W.W. 1994. Proceedings ICALEO '94, 77.

Gu, H., Duley, W.W. 1996a. *J. Phys. D.: Appl. Phys.* 29:550.

Gu, H., Duley, W.W. 1996b. *J. Phys. D.: Appl. Phys.* 29:556.

Gu, H., Duley, W.W. 1996c. Proceedings ICALEO '96, B40.

Guglielmino, E., LaRosa, G., Oliveri, S.M., et al. 1993. Proceedings ISATA '93, 57.

Habenicht, G., Stark, W., Deimann, R. 1991. *Weld. Cutting* 10:37.

Hall, B.E., Wallach, E.R. 1989. Recent Trends in Welding Science and Technology TWR '89, 775.

Hamann, C., Rosen, H.G., Lassiger, B. 1989. Proceedings SPIE 1132, 275.

Hanicke, L., Strandberg, O. 1993. SAE Int. Congr. and Exposition, Detroit, SAE Tech. Paper 93 0028, March.

Hansen, F., Duley, W.W. 1994. *J. Laser Appl.* 6:137.

Hanting, J., Aiquing, D. Proceedings LAMP '92, 457.

Haran, F.M., Hand, D.P., Peters, C., et al. 1996. Proceedings ICALEO '96, B49.

Haruta, K. 1995. Proceedings ICALEO '95, 866.

Harvey, M., Wallach, E.R. 1992. International Trends in Welding Science and Technology '92, 369.

Hashimoto, K., Sato, T., Niwa, K. 1991. *J. Laser Appl.* Winter: 21.

Hayashi, T., Inaba, Y., Matsuhiro, Y., et al. 1996. Proceedings ICALEO '96.

Heidecker, E., Schafer, J.H., Uhlenbusch, J., et al. 1988. *J. Appl. Phys.* 64:2291.

Heiple, C.R., Roper, J.R. 1982. *Weld. J.* 61:97s.

Heiple, C.R., Roper, J.R., Stagner, R.T., et al. 1983. *Weld. J.* March: 72s.

Hertzler, C., Wollermann-Windgasse, R. 1994. Proceedings SPIE 2206, 176.

Herziger, G. 1986. Proceedings SPIE 650, 188.

Herziger, G., Kreutz, E.W., Wissenbach, K. 1986. Proceedings SPIE 668, 2.

Heyden, J., Nilsson, K., Magnusson, C. 1990. *Ind. Laser Handbook,* 161.

Heyden, J., Nilsson, K., Magnusson, C. 1989. Proceedings 6th Int. Conf. Lasers in Manufacturing, 93.

Heyden, J., Nilsson, K., Magnusson, C. 1988. Proceedings LIM, 93.

Heyn, H., Decker, I., Wohlfahrt, H. 1994. Proceedings SPIE 2207, 381.

Hinricks, J.F., Ramsey, P.W., Ciaffoni, R.I., et al. 1974. *Weld. J.* 53:488.

Hirose, A., Arata, Y., Kobayashi, K.F. 1995. *J. Maternal Sci.* 30:970.

Hishii, M., Sato, K., Fukushima, T. 1987. Proceedings ICALEO '87, 109.

Honeycombe, J., Gooch, T.G. 1986. *Met. Constr.* Nov.: 703R.

Hongo, A., Morosawa, K., Shiota, T., et al. 1991. *Appl. Phys. Lett.* 58:1582.

Houlot, M., Nava-Rudiger, E. 1995. Proceedings ICALEO '95, 563.

Hoult, T. 1990. Proceedings SPIE 1277:209.

Huang, Q., Kullberg, G., Guan, Z., et al. 1993. Proceedings ISATA '93, 139.

Hügel, H. 1987. Proceedings 6th Int'l Symp. Gas Flow Chem. Lasers, Springer-Verlag, New York, 258.

Hügel, H., Dausinger, F., Berger, P., et al. 1994. Proceedings ECLAT, s63.

Hügel, H., Wiedmaier, M., Rudlaff, T. 1995. *Optical and Quant. Electron* 27:1149.

Hunter, B.V., Leong, K.H., Miller, C.B., et al. 1996. Proceedings ICALEO '96, E173.

Imhoff, R., Behler, K., Gatzweiler, W., et al. 1988. Proceedings 5th Int. Conf. Lasers in Manufacturing, 247.

Ion, J.C., Easterling, K.E., Ashby, M.F. 1984. *Acta. Metal.* 32:1949.

Ion, J.C., Salminen, A., Sun, Z. 1996. *Weld. J.* 75:225s.

Irving, B. 1991. *Weld. J.* 70, Sept.: 39.

Isenor, N.R. 1977. *Appl. Phys. Lett.* 31:148.

Ishide, T., Matsumoto, O., Nagura, Y., et al. 1990. Proceedings SPIE 1277:188.

Ishide, T., Shono, S., Ohmae, T., et al. 1987. Proceedings LAMP '87, Osaka, 187.

Ishihara, K., Troy, T.J., Shirai, K. 1993. Proceedings ICALEO '93, 435.

Iwai, Y., Okumura, N., Miyata, O. 1987. Proceedings LAMP '87, 517.

Iwamoto, H., Ebata, K., Namba, H. 1992. Proceedings LAMP, Osaka, 191.

Jain, A.K., Kulkarni, V.N., Sood, D.K., et al. 1981. *J. Appl. Phys.* 52:4882.

Jaroni, U., Pircher, H., Stegemann, T., et al. 1996. IBEC '96, Materials and Body Testing, 114.

Javan, A., Bennett, W.R., Herriott, D.R. 1961. *Phys. Rev. Lett.* 6:106.

Jellison, J.L., Keicher, D.M., Fuerschbach, P.W. 1990. Proceedings ICALEO '90, 123.

Jon, M.C. 1985. *Weld. J.* 64:43.

Jones, I., Riches, S., Yoon, J.W., et al. 1992. Proceedings LAMP '92, 523.

Junai, A.A., van Dijk, M., Hiensch, M., et al. 1990. Proceedings SPIE 1277, 217.

Jurca, M., Mokler, D., Ruican, R., et al. 1994. Proceedings SPIE 2207, 342.

Kaierle, S., Dahmen, M., Furst, B., et al. 1996. Proceedings ICALEO '96, B154.

Kalberer, M., Mueller, R., Sharp, M., McCay, M.H. 1994, Proceedings ICALEO '94, p. 732.

Kamalu, J.N., McDarmaid, D., Steen, W.M. 1991. Proceedings ICALEO '91, SPIE 1722: 178.

Kannatey-Asibu, E. 1989. Recent Trends in Welding Science and Technology TWR '89, 443.

Kannatey-Asibu, E. 1991. *ASME J. Eng. Mat.* 113:915.

Kapadia, P., Ducharme, R., Dowden, J. 1991. Proceedings ICALEO '91, SPIE 1722:53.

Kaplan, A. 1994. *J. Phys. D.: Appl. Phys.* 27:1805.

Kar, A., Mazumder, J. 1995. *J. Appl. Phys.* 78:6353.

Kar, A., Rockstroh, T., Mazumder, J. 1992. *J. Appl. Phys.* 71:2560.

Katayama, S., Kohsaka, S., Mizutami, M., et al. 1993. Proceedings ICALEO '93, 487.

Katayama, S., Lundin, C.D., Danko, J.C., et al. 1989. Recent Trends in Welding Science and Technology TWR '89, 687.

Kattamis, T.Z. 1981. Proceedings LIM, 1.

Kawali, S.M., Vieglahn, G.L., Scheuerman, R. 1991. Proceedings ICALEO '91, SPIE 1722: 156.

Kaye, A.S., Delph, A.G., Hanley, E., et al. 1983. *Appl. Phys. Lett.* 43:412.

Keicher, D.M., Essien, M. 1995. Proceedings ICALEO '95, 1092.

Keilmann, F. 1983. *Phys. Rev. Lett.* 51:2097.

Keilmann, F., Bai, Y.H. 1982. *Appl. Phys.* A29:9.

Kerby, F. 1964. U.S. Patent 3159419.

Khan, P.A. 1993. Proceedings ICALEO '93, 805.

Khan, P.A.A., DebRoy, T., David, S.A. 1988. *Weld. J.* 67:1s.

Kim, J.S., Watanabe, T., Yoshida, Y. 1993. Proceedings ICALEO '93, 640.

Kimara, S., Sugiyama, S., Mizutame, M. 1986. Proceedings ICALEO '86, 89.

Kinsman, G., Duley, W.W. 1993. Proceedings ICALEO '93.

Kinsman, G., Duley, W.W. 1986. Proceedings SPIE 668:19.

Kitani, Y., Yasuda, K., Kataoka, Y. 1995. Proceedings ICALEO '95, 514.

Klein, T., Vicanek, M., Kroos, J., et al. 1994. *J. Phys. D.: Appl. Phys.* 27:2023.

Klein, T., Vicanek, M., Simon, G. 1996. *J. Phys. D.: Appl. Phys.* 29:322.

Klemans, P.G. 1976. *J. Appl. Phys.* 47:2165.

Kluft, W., Boerger, P., Schwartz, R. 1996. Proceedings ICALEO '96, auto, 63.

Knight, C.J. 1979. *AIAA J.* 17:519.

Knudtson, J.T., Green, W.B., Sutton, D.G. 1987. *J. Appl. Phys.* 61:4771.

Koechner, W. 1992. Solid State Laser Engineering, 3rd ed., Springer-Verlag, New York.

Konstantinov, S., Smurov, I., Flamant, G. 1994. Proceedings ICALEO '94, 684.

Koons, J.N., Roessler, D.M. 1994. Proceedings IBEC '94, Advanced Technologies and Processes, 97.

Kosuge, S., Ono, M., Nakada, K., et al. 1986. Proceedings ICALEO '86, 105.

Kou, S., Hsu, S.C., Mehrabian, R. 1981. *Met. Trans.* 12B:33.

Kou, S., Wang, Y.H. 1986. *Met. Trans.* 17A:2265.

Kreidler, V., Nitsch, H. 1993. Proceedings ISATA '93, 161.

Kreutz, E.W., Ollier, B., Pirch, N. 1992. Proceedings LAMP, Osaka, 353.

Kroos, J., Gratzke, U., Simon, G. 1993a. *J. Phys. D.: Appl. Phys.* 26:474.

Kroos, J., Gratzke, U., Vicanek, M., et al. 1993b. *J. Phys. D.: Appl. Phys.* 26:481.

Kugler, T.R., Bransch, H. 1993. Proceedings ICALEO '93, 814.

Kuhnert, K.D. 1993. Proceedings ISATA '93, 203.

Kujanpaa, V.P., David, S.A. 1986. Proceedings ICALEO '86, 63.

Kujanpaa, V.P., Helin, J.P., Moisio, T.J.I. 1988. Proceedings ICALEO '88, 259.

Kujanpaa, V.P., Moisio, T.J.I. 1989. Recent International Trends in Welding Science and Technology TWR '89, 333.

Kurz, W., Giovanola, B., Trivedi, R. 1986. *Acta Metall.* 34:823.

Lacroix, D., Jeandel, G., Boudot, C. 1997. *J. Appl. Phys.* 81:6599.

Laflamme, G. 1996. Proceedings IBEC '96, Materials and Body Testing, 117.

Laflamme, G., Schubert, G. 1996. IBEC '96, Materials and Body Testing, 117.

Laflamme, G.R., Powers, D.E. 1987. Proceedings Conf. Laser vs. Electron Beam, Bakish Mat'l. Corp., 168.

Lambrakos, S.G., Metzbower, E.A., Moore, P.G., et al. 1991. Proceedings ICALEO '91, SPIE 1722:40.

Lampa, C., Powell, J., Ivarson, A., et al. 1995. Proceedings ICALEO '95, 504.

Lancaster, J.F. 1993. Metallurgy of Welding, 5th ed., Chapman & Hall, London.

Larsson, J.K. 1993. Proceedings ISATA '93, 29.

Ledenev, V.I., Mirzoyer, F.K., Nikolo, V.A. 1994. Proceedings SPIE 2257:10.

Leong, K.H., Holdridge, D.J., Liu, Y., et al. 1993. Proceedings ICALEO '93, 704.

Li, L., Qi, N., Brookfield, D.J., et al. 1990. Proceedings ICALEO '90, SPIE 1601:411.

Li, L., Steen, W.M. 1992. Proceedings ICALEO '92, 719.

Li, L., Steen, W.M., Modern, P. 1993. Proceedings ICALEO '93, 372.

Liang, Y., Zhang, B., Feng, Z. 1992. Proceedings LAMP, Osaka, 547.

Lim, G.C., Steen, W.M. 1984. *J. Phys. E. Sci. Instrum.* 17:999.

Lingenfelter, A.C. 1987. Proceedings LAMP '87, 211.

Liu, P.S., Baeslack, W.A., Hurley, J. 1994. *Weld. J.* 73:175s.

Liu, Y.N., Kannatey-Asibu. 1993. *Trans. ASME* 115:34.

Locke, E.V. 1972. *Weld. J.* 51:245s.

Locke, E.V., Hella, R.A. 1974. *IEEE J. Quant. Electron.* QE-10:179.

Locke, E.V., Hoag, E.D., Hella, R.A. 1972a. *IEEE J. Quant. Electron.* QE-8:132.

Locke, E., Hoag, E., Hella, R. 1972b. *Weld. J.* 51:245s.

Loosen, P. 1992. Proceedings SPIE 1810:26.

Lucas, J., Smith, J.S. 1988. Proceedings SPIE 952:559.

Macken, J. 1992. Proceedings LAMP '92, 67.

Magee, K.H., Merchant, V.E., Hyatt, C.V. 1990. Proceedings ICALEO '90, 382.

Maiman, T.H. 1960. *Nature* 187:493.

Maischner, D., Drenker, A., Seidel, B., et al. 1991. Proceedings ICALEO '91, 150.

Maisenhalder, F., Chen, G., Roth, G. 1993. Proceedings ISATA '93, 335.

Maiwa, T., Miyamoto, I., Mori, K. 1995. Proceedings ICALEO '95, 708.

Mallory, L., Orr, R.F., Wells, W. 1988. Laser Materials Processing III, Mazumder, J., Mukherjee, K.N., eds. *TMS* 123.

Manes, K.R., Zapata, L.E. 1990. Proceedings ICALEO '90, SPIE 1601:100.

Mannik, L., Brown, S.K. 1990. Proceedings ICALEO '90, SPIE 1601:364.

Mao, Y.L. 1993. Masters Thesis. University of Waterloo.

Mao, Y.L., Kinsman, G., Duley, W.W. 1993. *J. Laser Appl.* 5:17.

Marcatili, E.A.J., Schmeltzer, R.A. 1964. *Bell Syst. Tech. J.* 43:1783.

Marinoni, G., Maccagno, A., Rabino, E. 1989. Proceedings 6th Int. Conf. Lasers in Manufacturing, 106.

Marsico, T.A., Denney, P.E., Furio, A. 1993. Proceedings ICALEO '93, 445.

Marsico, T.A., Kossowsky, R. 1988. Laser Mater. Proceedings III, 135.

Martukanitz, R.P., Howell, P.R., Pratt, W.A. 1992. International Trends in Welding Science and Technology '92, 271.

Martukanitz, R.P., Smith, D.J., Armao, F.G., et al. 1994. SAE Tech. Paper 940158.

Maruo, H., Miyamoto, I., Arata, Y. 1985. IIW Doc. IV-391-85, Dept. Weld. Eng., Osaka University.

Maruo, H., Miyamoto, I., Arata, Y. 1981. Proceedings ICALEO '81, 98.

Matsuda, J., Utsumi, A., Katsumura, M., et al. 1988. *Join. Mat.* July: 31.

Matsuhiro, Y., Inaba, Y., Ohji, T. 1992. Proceedings LAMP '92, Osaka, 381.

Matsumura, H., Orihashi, T., Nakayama, S., et al. 1992. Proceedings LAMP '92, 529.

Matsumoto, T., Fukuda, N., Kondo, Y., et al. 1996. Proceedings ICALEO '96, E163.

Matsumoto, T., Seki, Y., Yasuda, K. 1992. Proceedings LAMP, Osaka, 225.

Matsunawa, A. 1990. Proceedings ICALEO '90, SPIE 1601:313.

Matsunawa, A. 1994. Proceedings ICALEO '94, 203.

Matsunawa, A., Katayama, S., Ikeda, H., et al. 1992. Proceedings ICALEO '92, 547.

Matsunawa, A., Kim, J.D., Takemoto, T., et al. 1995. Proceedings ICALEO '95, 719.

Matsunawa, A., Kim, J.D., Katayama, S., et al. 1996. Proceedings ICALEO '96, B58.

Mazumder, J. 1983. Lasers for Materials Processing, Bass, M., ed., North-Holland Publ. Co., New York, 113.

Mazumder, J. 1991. *Opt. Eng.* 30:1208.

Mazumder, J., Rockstroh, T.J., Krier, H. 1987. *J. Appl. Phys.* 62:4712.

Mazumder, J., Steen, W.M. 1979. TMS-AISE Fall Mtg., Paper F79-17, Sept.

Mazumder, J., Steen, W.M. 1980a. *J. Appl. Phys.* 51:941.

Mazumder, J., Steen, W.M. 1980b. *Met. Constr.,* Sept., 423.

Mazumder, J., Voekel, D. 1992. Proceedings LAMP, 373.

McCay, M.H., McCay, T.D., Dahotre, N.B., et al. 1991. *J. Laser Appl.* 3:35.

McCay, M.H., McCay, T.D., Sedghinasab, A., et al. 1989. Laser Materials Proceedings III, Mazumder, J., Mukherjee, K.N., eds. Minerals, Metals, & Materials Soc., 107.

McCay, M.H., Sharp, C.M., Womacky, M.G., et al. 1990. Proceedings ICALEO '90, SPIE 1601:325.

McIver, J.K., Guenther, A.H. 1986. Proceedings SPIE 650:123.

Mehmetli, B., Takahoshi, K., Sato, S. 1996. *J. Laser Appl.* 8:25.

Metzbower, E.A. 1983. *Met. Constr.,* Oct.:611.

Metzbower, E.A. 1990. *Weld. J.* 69:272s.

Metzbower, E.A. 1992. Proceedings ICALEO '92, 163.

Metzbower, E.A. 1993a. *Met. Trans.* 24B:875.

Metzbower, E.A. 1993b. *Weld. J.* 72:403s.

Metzbower, E.A., DeMarco, L., Denney, P.E., et al. 1993. Proceedings ICALEO '93, 153.

Metzbower, E.A., Denney, P.E., Pratt, W., et al. 1994. Proceedings ICALEO '94, 846.

Metzbower, E.A., Moon, D.W. 1981. Proceedings LIM, 255.

Michelle, A., Schafer, J.H., Uhlenbusch, J., Viol, W. 1990. Proceedings SPIE 1276:231.

Milewski, J.O., Keel, G., Sklar, E. 1995. Proceedings ICALEO '95, 875.

Milewski, J.O., Lewis, G.K., Wittig, J.E. 1993. *Weld. J.* 72:341s.

Miller, K.J., Nunnikhoven, J.D. 1965. *Weld. J.* 44:480.

Minamida, K., Sugihashi, A., Kido, M., et al. 1991. Proceedings ICALEO '91, SPIE 1722: 168.

Minamida, K., Takafuji, H., Hamada, N., et al. 1986. Proceedings ICALEO '86, 97.

Miura, H., Shibano, I. 1990. *Trans. J. Weld. Soc.* 21:2.

Miyamoto, I., Maruo, H. 1992. Proceedings LAMP, Osaka, 311.

Miyamoto, I., Maruo, H., Kuriyama, K., et al. 1987. Proceedings LAMP '87, 231.

Miyamoto, I., Maruo, H., Arata, Y. 1984. Proceedings ICALEO '84, 313.

Miyamoto, I., Maruo, H., Arata, Y. 1986. Proceedings SPIE 668, 11.

Miyamoto, I., Kamimuki, K., Maruo, H., et al. 1993. Proceedings ICALEO '93, 413.

Miyamoto, I., Mori, K. 1995. Proceedings ICALEO '95, 759.

Miyamoto, I., Nanba, H., Maruo, H. 1990. Proceedings SPIE 1276:112.

Miyamoto, I., Uchida, T., Maruo, H., et al. 1994. Proceedings ICALEO '94, 293.

Miyata, T. 1986. Proceedings SPIE, 650:131.

Molian, P.A., Srivatsan, T.S. 1988. Proceedings ICALEO '88, 238.

Mombo-Caristan, J.C. 1996a. Automotive Laser Applications Workshop, University of Michigan (preprint).

Mombo-Caristan, J.C. 1996b. Proceedings ICALEO '96, D46.

Mombo-Caristan, J.C. 1997. Proceedings IBEC '97, Body Assembly and Manufacturing, 88.

Moon, D.W., Metzbower, E.A. 1983. *Weld. J.* 62:53s.

Morgan, S.A., Hand, D.P., Haran, F.M., et al. 1996. Proceedings ICALEO '96, D28.

Morochko, V.P., Fedorov, B.M., Andreev, V.D. 1983. *Auto Weld.* June: 13.

Morrow, C.E., Gu, G. 1993. Proceedings ICALEO '93, 403.

Mueller, R. 1994a. Proceedings ICALEO '94, 509.

Mueller, R.E. 1994b. Doctoral Thesis, York University.

Mueller, R.E., Hopkins, J.A., Semak, V.V., et al. 1996. ICALEO '96, B86.

Muller, M., Dausinger, F., Griebsch, J. 1997. Proceedings IBEC '97, Advanced Technologies and Processes, 62.

Mundra, K., DebRoy, T. 1992. Proceedings International Trends in Welding Science and Technology '92, 75.

Mundra, K., DebRoy, T. 1993a. *Weld. J.* 72:1s.

Mundra, K., DebRoy, T. 1993b. *Met. Trans.* 24B:145.

Nakajima, N., Shimokusu, Y., Shono, S., et al. 1989. *Weld. World* 27:43.

Narikiyo, T., Fujinaga, S., Miura, H., et al. 1995. Proceedings ICALEO '95, 885.

Natsumi, F., Ikemoto, K., Sugiura, H., et al. 1992. *Int. J. Materials Prod. Technol.* 7:193.

Newsome, P.A., Meyer, D.W., Albright, C.A. 1987. Proceedings LAMP '87, 193.

Nilsson, K., Magnusson, C., Tapper, L., et al. 1993. Proceedings ISATA '93, 73.

Nishimi, A., Kanazawa, H., Taniu, Y., et al. 1996. Proceedings ICALEO '96, E11.

Nonhof, C.J. 1988. Material Processing with Nd: YAG Lasers, Electrochemical Publications, Ayr, Scotland.

Nonhof, C.J. 1994. *Polym. Eng. Sci.* 34:1547.

Norris, I., Hoult, T., Peters, C., et al. 1992. Proceedings LAMP '92, 489.

Notenboom, G., Nonhof, C., Schildbach, K. 1987. Proceedings LAMP '87, 107.

Ocana, J.J., Herrero, F., Chaya, C., et al. 1994. Proceedings SPIE 2207, 428.

Ohji, T., Murakami, E., Matsubatashi, K., et al. 1994. Proceedings ICALEO '94, 471.

Ohji, T., Shiwaka, T., Kimura, K., et al. 1995. Proceedings ICALEO '95, 729.

Oikawa, M., Minamida, K., Goto, N., et al. 1993. Proceedings ICALEO '93, 453.

Olfert, M. 1998. Doctoral Thesis, University of Waterloo.

Olfert, M., Bridger, P., Dunn, I., et al. 1994. *High Temp. Chem. Processes* 3:1.

Olfert, M., Duley, W.W. 1996. Proceedings ICALEO '96, B146.

Olsen, F.O. 1980. *DVS Ber.* 63:197.

Olsen, F.O. 1994. Proceedings SPIE 2207:27.

Olsen, F.O., Jorgensen, H., Bagger, C., et al. 1992. Proceedings LAMP '92, 405.

O'Neill, W., Steen, W.M. 1988. Proceedings ICALEO '88, 90.

Otto, A., Deinzer, G., Geiger, M. 1994. Proceedings SPIE 2207:282.

Otto, A., Geisel, M., Geiger, M. 1996. Proceedings ICALEO '96, B30.

Paul, A., DebRoy, T. 1988. *Met. Trans.* 19B:851.

Paul, A.J., Khan, P.A. 1994. *J. Laser Appl.* 6:32.

Penasa, M., Columbo, E., Giolfo, M. 1994. Proceedings SPIE 2207:200.

Petrick, F.D. 1990. U.S. Patent 4916284.

Petrolonis, K. 1993. *Weld. J.* 72:301s.

Pfluger, A.R., Maas, P.M. 1965. *Weld. J.* 44:264.

Phillips, R.H., Metzbower, E.A. 1992. *Weld. J.* 71:201s.

Phitzer, E.K., Turner, R. 1968. *J. Sci. Instrum. (J. Phys. E.)* 1:360.

Piane, A.D., Sartorio, F., Cantetto, M., et al. 1987. U.S. Patent 4682002.

Pickering, E.R., Giagola, M.A., Ramage, R.M., et al. 1995. Aluminum Applications for Automotive Design, SAE, SP1097, 67.

Pirch, N., Kreutz, E.W., Ollier, B., et al. 1996. Proceedings NATO ASI E307, 177.

Pitscheneder, W., DebRoy, T., Mundra, K., et al. 1996. *Weld. J.* 75:71s.

Polzin, M., Poprawe, R., Kawalla, R., et al. 1995. Proceedings IBEC '95, Advanced Technologies and Processes, 124.

Poueyo, A., Deshors, G., Febbro, R., et al. 1992. Proceedings LAMP, 323.

Poueyo-Verwaerde, A., Fabbro, R., Deshors, G., et al. 1993. *J. Appl. Phys.* 74:5773.

Postacioglu, N., Kapadia, P., Davis, M., et al. 1987. *J. Phys. D.: Appl. Phys.* 20:340.

Postacioglu, N., Kapadia, P., Dowden, J. 1989. *J. Phys. D.: Appl. Phys.* 22:1050.

Postacioglu, N., Kapadia, P., Dowden, J. 1991. *J. Phys. D.: Appl. Phys.* 24:15.

Rajendra, N., Pate, M.B. 1988. Proceedings ICALEO '88, 119.

Rapp, J., Glumann, C., Dausinger, F., et al. 1993. Proceedings ISATA '93, 95.

Rapp, J., Glumann, C., Dausinger, F., et al. 1995. *Optical Quant. Electron* 27:1203.

Rappaz, M., David, S.A., Vitek, J.M., et al. 1989. Recent International Trends in Welding Science and Technology TWR '89, 147.

Rath, W., Northemann, T. 1994. Proceedings SPIE 2206, 185.

Ready, J.F. 1971. Effects of High Power Laser Radiation, Academic Press, New York.

Richter, K., Eberle, H.G., Maucher, K.H. 1993. Proceedings ISATA '93, 103.

Rito, N., Ohta, M., Yamada, T., et al. 1988. U.S. Patent 4745257.

Rockstroh, T.J., Mazumder, J. 1987. *J. Appl. Phys.* 61:917.

Rosenthal, D. 1976. *Trans. Amer. Soc. Mech. Eng.* 68:849.

Ruffler, C., Gürs, K. 1972. *Opt. Laser Technol.* 4:265.

Rykalin, N.N., Kransulin, Y.L. 1965. *Sov. Phys. Dokl.* 10:659.

Rykalin, N.N., Uglov, A.A. 1965. *Weld. Prod.* June: 14.

Rykalin, N., Uglov, A., Kokora, A. 1978. Laser Machining and Welding, MIR, Moscow.

Sahoo, P., Collur, M.M., DebRoy, T. 1988. *Met. Trans.* 19B:967.

Saifi, M.A., Vahaviolos, S.J. 1976. *IEEE J. Quant. Electron.* QE 12:129.

Sajatovic, N. 1996. Proceedings IBEC, Materials and Body Testing, 106.

Sakamoto, H., Shibata, K., Dausinger, F. 1992. Proceedings ICALEO '92.

Salminen, A., Kujanpaa, V.P., Moisio, T.J.I. 1994. Proceedings ICALEO '94, 193.

Salminen, A.S., Kujanpaa, V.P., Moisio, T.J.I. 1996. *Weld. J.* 75:9s.

Sarady, I., Lundquist, B., Magnusson, C. 1993. Proceedings ICALEO '93, 630.

Sasnett, M.W., Hurley, J.P. 1994. Proceedings ICALEO '94, 742.

Sato, S., Takahashi, K., Mehmetli, B. 1996. *J. Appl. Phys.* 79:8917.

Saunders, F.I., Wagoner, R.H. 1996. *Met. Mat. Trans. A.* 27A:2605.

Scheurman, R. 1993. Proceedings ISATA '93, 89.

Schimon, R.W., Mazumder, J. 1993. Proceedings ICALEO '93, 382.

Schmidt, A.O., Ham, I., Hoshi, T. 1965. *Weld. J.* 44:481.

Schmitz, B., Defourny, J. 1992. *Intl. Trends in Welding Sci. and Technol.* 427.

Schou, C.E., Semak, V.V., McCay, T.D. 1994. Proceedings ICALEO '94, 41.

Schraft, R.D., Hardock, G., Konig, M. 1988. Proceedings SPIE 952, 552.

Schulz, W., Behler, K. 1989. Proceedings SPIE 1132:166.

Schulz, W., Furst, B., Kaierle, S., et al. 1996. Proceedings ICALEO '96, D1.

Schuöcker, D. 1991. Proceedings ICALEO '91, SPIE 1722:32.

Schuöcker, D., Kaplan, A. 1994. Proceedings SPIE 2207:236.

Sepold, G., Zierau, M. 1992. Proceedings LAMP '92, 219.

Seidel, B., Beersiek, J., Beyer, E. 1994. Proceedings SPIE 2207:279.

Seidel, B., Sokolowski, W., Beersiek, J., et al. 1993. Proceedings ISATA '93, 187.

Seiler, P. 1988. Proceedings 5th Int. Conf. Lasers Manufacturing, 157.

Sekhar, J.A., Kou, S., Mehralian, R. 1983. *Met. Trans.* 14A:1169.

Semak, V.V., Hopkins, J.A., McCay, M.H., McCay, T.D. 1994a, Proceedings ICALEO '94, 641.

Semak, V.V., Hopkins, J.A., McCay, M.H., et al. 1994b. Proceedings ICALEO '94, 830.

Semak, V.V., Hopkins, J.A., McCay, M.H. et al. 1995a. Proceedings ICALEO '95, 739.

Semak, V.V., McCay, M.H., McCay, T.D. 1993. Proceedings ICALEO '93, 777.

Semak, V.V., West, J.C., Hopkins, J.A., et al. 1995b. Proceedings ICALEO '95, 544.

Sepold, G. 1984. *Weld. Cutting* May: 30.

Sepold, G., Rothe, R., Teske, K. 1987. Proceedings LAMP '87, 151.

Shannon, G.J., Nuttall, R., Watson, J., et al. 1994a. *Weld. J.* 73:173s.

Shannon, G.J., Watson, J., Deans, W.F. 1994b. *J. Laser Appl.* 6:223.

Shaw, H.F. 1994. TWI Technology Brief, 488/1994.

Sherman, G.H., Danielewicz, E.J., Rudisill, J.E., et al. 1987. Proceedings LAMP, Osaka, 625.

Shewell, J.R. 1977. *Weld. Design Fabr.* June: 106.

Shi, K., Li, L., Steen, W.M., et al. 1992. Proceedings LAMP '92, 451.

Shibata, K. 1996. Proceedings Conf. Lasers and Electrooptics for Automotive Manufacturing, LIA, 11.

Shida, T., Hirokawa, M., Sato, S. 1997. *Q. J. Jpn. Weld. Soc.* 15:18.

Shimbo, Y., Ono, M., Kabasawa, M. 1993. Proceedings ICALEO '93, 712.

Shinoda, T., Matsunaga, K., Akaishi, T. 1992. Proceedings LAMP, Osaka, 541.

Siekmann, J., Morijn, R. 1968. *Phillip. Res. Rep.* 23:367.

Simidzu, H., Yoshiro, F., Katayama, S., et al. 1992. Proceedings LAMP '92, 511.

Simon, G., Gratzke, U., Kroos, J. 1993. *J. Phys. D.: Appl. Phys.* 862.

Smith, J.F., Thompson, A. 1967. Rec. IEEE Symp. Electron Ion and Laser Beam Technology, 268.

Sokolowski, W., Herziger, G., Beyer, E. 1989. Proceedings SPIE 1132:288.

Spies, B., Thomas, V. 1992. U.S. Patent 5104032.

Spitzer, L. 1968. Physics of Fully Ionized Gases, Interscience, New York.

Stares, I.J., Apps, R.L., Megaw, J.H.P.C., et al. 1987. *Met. Constr.* March: 27.

Starzer, M., Ebner, R., Glatz, W., et al. 1993. Proceedings ISATA '93, 131.

Steen, W.M. 1986. Industrial Laser Annual Handbook, Belforte, D., Levitt, M., eds. Pennwell Books, Tulsa, 158.

Steen, W.M. 1991. Laser Material Processing. Springer-Verlag, New York.

Steen, W.M. 1992. Proceedings LAMP '92, 439.

Steen, W.M. 1980. *J. Appl. Phys.* 51:5636.

Steen, W.M., Chen, Z.D., West, D.R.F. 1987. Industrial Laser Handbook, Belforte, D., Levitt, M. eds. Penwell Books, Tulsa, 80.

Steen, W.M., Eboo, M. 1979. *Met. Constr.* July: 332.

Steen, W.M., Li, L. 1988. Proceedings SPIE 952:544.

Steen, W.M., Weerasinghe, V.M. 1986. Proceedings SPIE 668:37.

Stoop, J., Metzbower, E.A. 1978. *Weld. J.* 57:345.

Sugawara, H., Kuwabara, K., Takemori, S., et al. 1987. Proceedings 6th Int'l. Symp. Gas Flow Chem. Lasers, Springer-Verlag, New York, 265.

Sun, Z., Ion, J.C. 1995. *J. Mat'l. Sci.* 30:4205.

Sun, Z., Salminen, A.S., Moisio, T.J.I. 1993. *J. Matl. Sci. Lett.* 12:1131.

Swift-Hook, D.T., Gick, A.E.F. 1973. *Weld. J.* 52:492s.

Szymanski, Z., Kurzyna, J. 1994. *J. Appl. Phys.* 76:7750.

Takano, G., Matsumoto, O., Nagura, Y., et al. 1990. Proceedings ICALEO '90, 373.

Terasaki, T. 1981. *J. Iron Steel Inst. Japan* 16:145.

Thorstensen, B. 1989. *Ind. Laser Handbook.* 64.

Tix, C., Simon, G. 1993. *J. Phys. D.: Appl. Phys.* 26:2066.

Tix, C., Simon, G. 1994. *Phys. Rev. E.* 50:453.

Tönshoff, H.K., Byun, C.W. 1992. Proceedings LAMP '92, 367.

Tönshoff, H.K., Gonschior, M. 1993. Proceedings ICALEO '93, 507.

Tönshoff, H.K., Meyer-Kobbe, C., Beske, E. 1990. Proceedings SPIE 1277:199.

Tönshoff, H.K., Overmeyer, L., Schumacker, J. 1996. Proceedings ICALEO '96, auto, 45.

Trappe, J., Kroos, J., Tix, C., et al. 1994. *J. Phys. D.: Appl. Phys.* 27:2152.

Trunzer, W., Lindl, H., Schwarz, H. 1993. Proceedings ISATA '93, 195.

Tsukamoto, S., Hiraoka, K., Asai, Y., et al. 1996. Proceedings ICALEO '96, B76.

Uddin, N., Beradi, E., DuCharme, R.C., et al. 1986. Proceedings SPIE 668:260.

Uddin, N., Watt, D.F. 1994. Proceedings ICALEO '94, 443.

Ursu, I., Mihailescu, I.N., Popa, A., et al. 1984. *Appl. Phys. Lett.* 45:365.

Ursu, I., Mihailescu, I.N., Prokhorov, A.M., et al. 1987. *J. Appl. Phys.* 61:2445.

van der Hoeven, J.M., Mohrbacher, H., Rubben, K., et al. 1996. IBEC '96, Materials and Body Testing, 95.

Vastra, I., Diotalevi, M. 1994. Proceedings SPIE 2207:439.

Venkat, S., Albright, C.E., Ramasamy, S., et al. 1997. *Weld. J.* 76:275-s.

Walther, H.W. 1994. IBEC '94. Advanced Technologies & Processes, 92.

Verwaerde, A., Fabbro, R., Deshors, G. *J. Appl. Phys.* 78, 2981, 1995.

Vilar, R., Miranda, R.M. 1988. Proceedings SPIE 952, 719.

Vitek, J.M., David, S.A. 1994. *Proceedings Laser Matl. Processes* IV, eds. Mazumder, J., Mukherjee, K., Mordike, B.L., Minerals, Metals & Materials Soc., 153.

von Allmen, M. 1976. *J. Appl. Metals* 47, 5460.

Waddell, W., Williams, N.T., Haberfield, A.B. 1987. *Met. Constr.* June: 313.

Walther, H.W. 1994. Proceedings IBEC '94. Advanced Technologies and Processes, 92.

Wang, P.C. 1993. *Weld. J.* 72, 155s.

Wang, P.C., Davidson, S. 1992. *Weld. J.* 71, 209s.

Wang, P.C., Ewing, K.M. 1991. *Weld. J.* 70, 43.

Wang, P.C., Ewing, K.M. 1994a. *Weld. J.* 73, 209s.

Wang, P.C., Ewing, K.M. 1992b. *J. Laser Appl.* 6, 14.

Wang, S.C., Wei, P.S. 1992. *Met. Trans.* 23B, 505.

Watanabe, M., Okado, H., Inoue, T., Nakamura, S., Matunawa, A. 1995. Proceedings ICALEO. '95, 553.

Watanabe, T., Yoshida, Y. 1990. *JSME Intl. J.* 33, 575.

Watanabe, T., Yoshida, Y. 1991. Model Casting, Welding Adv. Solidification Processes V, eds., Rappa, M., Ozgu, M.R., Mahin, K.W., Minerals, Metals & Materials Soc., 173.

Watson, M.N., Dawes, C.J. 1985. *Met. Constr.* Sept.: 561.

Watt, D.F., Uddin, M.N. 1994. Proceedings ICALEO '94, 67.

Webster, J.M. 1970. *Met. Progr.* 98:59.

Weedon, T.M.W. 1987. Proceedings LAMP '87, 75.

Weeter, L., Albright, C.E., Jones, W.H. 1986. *Weld. J.* 66:51-s.

Weiting, T.J., de Rosa, J.L. 1979. *J. Appl. Phys.* 50:1071.

Whitehouse, D.R., Nilsen, C.J. 1990. Proceedings ICALEO '90, SPIE 1601:13.

Wildmann, D., Urech, W., Freuler, E. 1996. IBEC '96, Materials and Body Testing, 132.

Willgoss, R.A., Megaw, J.H.P.C., Clark, J.N. 1979. *Opt. Laser Technol.* April: 73.

Williams, K., Steen, W.M., Ducharme, R., et al. 1993a. Proceedings ICALEO '93, 168.

Williams, S.W., Salter, P.L., Scott, G., et al. 1993b. Proceedings ISATA '93, 49.

Willmott, N.F.F., Hibbard, R., Steen, W.M. 1988. Proceedings ICALEO '88, 109.

Wojcicki, M.A., Pryputniewicz, R.J. 1996. Proceedings ICALEO '96, D-84.

Wollermann-Windgasse, R., Ackerman, F., Weick, J., et al. 1986. Proceedings ICALEO '86, 45.

Woolin, P. 1994. *Weld. Mat. Fab.* Jan.: 18.

Xie, J., Kar, A., Rotherflue, J.A., et al. 1997. *J. Laser Appl.* 9:77.

Xijing, X., Katayama, S., Matsunawa, A. 1997. *Weld. J.* 76:70-s.

Yagi, S., Kuzumoto, M., Ohtani, A. 1992. Proceedings LAMP '92, 91.

Yang, Y.S., Hsu, C.R., Albright, C.E., et al. 1993. *J. Laser Appl.* 5:17.

Yip, W.M., Man, H.C., Ip., W.H. 1996. Proceedings ICALEO '96, D74.

Yoshida, M., Yamasaki, Y., Kebasawa, S., et al. 1995. *NKK Tech. Rev.* 72:25.

Young, J.F., Preston, J.S., van Driel, H.M., et al. 1983. *Phys. Rev.* 27B:1155.

Yurioka, N., Okumura, M., Kasuva, T., et al. 1987. *Met. Constr.* 19:217R.

Zacharia, T., David, S.A., Vitek, J.M., et al. 1989. *Weld. J.* 68:499s.

Zerkle, D.K., Krier, H. 1994. *AIAA J.* 32:324.

Zimmerman, K. 1993. Proceedings ISATA '93, 171.

Zuyao, T., Chu, C., Bingyou, D. 1987. Proceedings LAMP, Osaka, 217.

# *Index*